THE WORLD'S WORST CARS

THE WORLD'S WORST CARS

FROM PIONEERING FAILURES TO MULTIMILLION DOLLAR DISASTERS

CRAIG CHEETHAM

This edition first published in 2005 for Grange Books
An imprint of Grange Books Ltd
The Grange
Kingsnorth Industrial Estate
Hoo, nr Rochester
Kent ME3 9ND
www.grangebooks.co.uk

Reprinted in 2006, 2007

ISBN-10: 1-84013-751-7
ISBN-13: 978-1-84013-751-4

Editorial and design by
Amber Books Ltd
Bradley's Close
74–77 White Lion Street
London N1 9PF
www.amberbooks.co.uk

Project Editor: Michael Spilling
Design: Hawes Design

Printed in Singapore

CONTENTS

INTRODUCTION

There are hundreds of books about the world's best cars. Some models even have entire volumes dedicated to them. But what about motoring's underclass? The cars that spent their lives as the butts of many a joke, suffered at the hands of stand-up comedians and earned their place in the annals of motoring history not for what they achieved, but for what they didn't. Despite their many and various faults, some of them even made a dedicated band of owners very happy.

This book is dedicated to the underdogs: the design disasters, financial failures and motoring misfits that make the motor industry such a fascinating and exciting field in which to work. It is in no way intended to insult and deride these machines. After all, a car may have become a spectacular failure after its launch, but you can be sure that, somewhere along the line, it was the dream of at least one person. The world also needs brilliant engineers, and those in the motor industry are, and always have been, among the global elite. This book intends not to slight their achievements, but to go some way to explaining why some of the world's least spectacular cars came into being.

Each of the 150 models has been chosen for a particular reason, and the opinions expressed are my own. There are many more cars out there that could easily qualify for selection, and you may feel that some of the cars

Above: *Leyland Australia marketed the P76 as 'Anything but average'. In reality, it wasn't even good enough to earn 'average' status.*

Above: ***AMC's Eagle may have set the precedent for the current trend towards car-based sports utility vehicles (SUVs), but it was an absolute horror to drive.***

appearing here do not deserve such criticism. Certainly, some of my choices are controversial, but I hope I explain myself fully in the accompanying text. There are even cars in here that I would happily jump to defend. I've owned at least a dozen of the cars in this book, and speak from experience when I say that my memory of owning them was nowhere near as bad as some of the reports they received when new.

I've also tried to make this book as global as possible. Sheer volume means that the majority of cars chosen are from the world's biggest car markets – the USA, Japan, Western Europe and Great Britain. But there are cars here from other countries, too: Russia, Iran, Australia, Poland and Korea are also represented here by their own transports of non-delight.

Above all, I want this book to be a tribute to and celebration of the diversity of the motor industry. A car may receive bad press when new, but it can still attract followers; without enthusiasts flying the flag for even the most awful cars ever made, the world would be a much duller place. To those souls brave enough to own, love and cherish one of the Worst Cars in the World, I salute your bravery and originality, and, inevitably, your sense of humour. Take a look at my driveway and you'll see I'm right there with you.

The 150 cars in this title have been split into five categories: Badly Built, Design Disasters, Financial Failures, Misplaced Marques and Motoring Misfits. A full explanation of the category is provided on the opening page of each section, but it's important to point out that in some cases the cars themselves were less to blame than the marketing departments.

ADVANCE SPECIFICATION

AUSTIN 3 LITRE

BADLY BUILT

Not all of the worst cars in the world were born bad. Some were brilliantly conceived and cleverly designed. In many cases, they could have been, and should have been, brilliant, but were let down by the very people that built them. Such was the case with the cars in this section. Some of them, including the Alfa Romeo Alfasud and Rover SD1, were brilliantly received on their debuts. The press hailed their design supremacy and, only later, after serious faults started to manifest themselves, did it become apparent that their owners were in for sleepless nights and financial hardship.

Others, of course, were complete duds from the outset. Cars such as the Austin Allegro and Renault 14 were not that desirable to start off with – and then shocking reliability records destroyed their last vestiges of respectability. Over the next few pages, we have detailed the absolute low points of the global motor industry. Not only were these cars some of the worst ever built, but also they were not even built properly to start with …

Left: *British Leyland's stillborn Austin 3-Litre was one of the most dreadful cars of its era, in every conceivable respect.*

ALFA ROMEO ALFASUD *(1972–84)*

The 'Sud was Europe's biggest motoring might-have-been. Brilliantly styled and fantastic to drive, it had the potential to become an all-time great and would have been just the tonic that the struggling Alfa Romeo needed to boost its sales charts. And so it nearly was. Initial reviews of the car were hugely positive, with road testers complimenting its fluent handling, lively flat-four engines and sporty nature. But the rot would soon set in – in the most literal sense. The 'Sud was so called because it was built in Southern Italy, at a new plant in Naples, where unemployment was rife.

Labour relations were dire, however, and cost-cutting led to the use of recycled Russian steel, which meant that the 'Sud was dreadfully made. Initially, it was the car's plastic trim that fell off – but, within two years, most 'Suds had started to rust like shipwrecks.

SPECIFICATIONS

TOP SPEED:	149km/h (93mph)
0–96KM/H (0–60MPH):	14.1secs
ENGINE TYPE:	flat-four
DISPLACEMENT:	1186cc (72ci)
WEIGHT:	823kg (1830lb)
MILEAGE:	7.6l/100km (37mpg)

Left: ***Alfa did its best when advertising the Sud, but although it was indeed a particularly good 'small package' to drive, the poor build quality became apparent all too soon.***

The 'Sud was neatly styled, but its looks faded quickly. So bad was the metal quality that rust soon spoiled the front wings, rear wheel arches and sills, and even appeared in the middle of panels, such as the roof and hood.

Alfa Romeo's flat-four 'Boxer' engines were renowned for their sporting character – and the Sud didn't disappoint. It was lively and sounded great, but reliability was let down by dire electrical problems.

The bodywork was rusty enough, but what lay underneath was even worse. After less than three years, most 'Suds had been extensively welded, such was their lack of structural integrity.

The great shame about the Alfasud was the fact that, build quality aside, it was a fabulous car. It was offered as a family saloon, but had the handling abilities of a genuine sports car.

ASTON MARTIN LAGONDA (1975–90)

Aston Martin already knew how to create cars that turned heads, but the Lagonda was something new and different from the company famous for its elegant but reserved styling. In true 1970s fashion, the four-door luxury saloon was wedge-shaped. And what a wedge it was! From bow to stern, the Lagonda was a dramatic if rather vulgar piece of styling, while inside it bristled with new electronic technology, including an all-LCD instrument pack. But sadly, it was underdeveloped, and electrical problems were rife.

The car's instrument packs gave up, the pop-up headlights stopped popping up and, to make matters worse, the floorpans tended to rot with surprising alacrity for what was such an expensive and flamboyant machine. Today, the Lagonda remains largely snubbed by Aston Martin fans, who much prefer the company's more traditional models.

SPECIFICATIONS

TOP SPEED:	231km/h (143mph)
0–96KM/H (0–60MPH):	8.8secs
ENGINE TYPE:	V8
DISPLACEMENT:	5340cc (326ci)
WEIGHT:	1984kg (4410lb)
MILEAGE:	20.1l/100km (14mpg)

Left: ***Not even the 'New for 1985' walnut dashboard, pepperpot alloy wheels and wood trim were enough to tempt more buyers into William Towns's controversial 'Wedge'.***

Aston Martin may have claimed that the Lagonda was bristling with new technology, but this did not extend under the hood. The wedge-shaped saloon came with the company's trademark 5.3-litre (323ci) V8 as standard.

The Lagonda's digital LCD dashboard was billed as a technological breakthrough when it first appeared, and other manufacturers did follow suit, but the British company's pioneering attempt at digital instruments was doomed to failure because the units broke down regularly.

Traditional Aston Martin fans did not take to the Lagonda's dramatic styling. It was certainly avant-garde, but few would call it genuinely stylish. It was too long and narrow, and the tail sat too high for the low nose.

It might have been expensive, but the Lagonda still had a build quality that was symptomatic of British models in the 1970s: corrosion was rife, especially in the sills and floor panels.

AUSTIN ALLEGRO *(1973–82)*

Possibly the most derided car in Europe, and certainly in Britain, the Austin Allegro was once British Leyland's great white hope. In reality, it became the British motor industry's white elephant. Not only did it look like an upturned bathtub, but it was also a disaster in terms of build quality, with panel gaps you could stick your arm through, trunks that let in water and windows that fell out if you jacked the car up in the wrong place. Throw in a square steering wheel – an eccentric idea in more senses than one – and, on plush models, a five-speed gearbox that felt like mixing concrete, and you have one of the biggest disasters in motoring history. The Allegro's case wasn't strengthened by the fact that it was built during an era of industrial unrest – many cars were left standing unfinished on the production line while the workers were out on yet another strike. Oddly, it has become a cult car, and there's now a global owners' club!

Left: ***These people aren't going diving. They're wearing wet suits only because their Allegro estate lets in so much water through its leaking window seals …***

SPECIFICATIONS

TOP SPEED:	82mph (132km/h)
0–96KM/H (0–60MPH):	18.4secs
ENGINE TYPE:	in-line four
DISPLACEMENT:	1275cc (78ci)
WEIGHT:	819kg (1822lb)
MILEAGE:	7.0l/100km (40mpg)

Entry-level models had the acceptable A-Series engine, but more powerful ones got the E-Series overhead cam unit, which came attached to a hideous gearbox, starved itself of oil and chewed its way through clutches.

The Allegro's steering wheel was supposedly revolutionary – but its revolutions were somewhat disorienting. The wheel was more square-shaped than round, for no other reason than to be different. After two years, it was scrapped.

Early in its design stages, the Allegro was a sleek, coupé-like creature. But the need to use the Maxi engine and heater, and to accommodate the Hydragas suspension, meant it ended up taller, wider and rounder. Or pig-ugly, to put it more succinctly.

Its complex suspension delivered a fairly compliant ride and negated the need for shock absorbers, but the pipes used to rust and leak fluid, causing the car to collapse on one side while parked overnight.

AUSTIN MAESTRO (1982–95)

Oh, dear! With the Allegro and Maxi finally put out to pasture, Austin-Rover was hoping for great things with the Maestro. But it wasn't to be. Once again, the stylists completely misjudged buyers' tastes, and the Maestro's tall stance and large glass areas quickly earned it the unfortunate nickname of 'Popemobile' from the British motoring press. Nor did things get any better as the car aged: there were terrible rust problems around the Maestro's rear wheel arches and a dated engine range offered little in the way of performance or refinement.

Top Vanden Plas models came with a speech synthesizer in the dashboard, voiced by New Zealand actress Nicolette Mackenzie. This feature frequently warned of low oil pressure and a lack of fuel, even when the car was in rude health, while it barked at you to put your seatbelt on before you'd even sat down.

SPECIFICATIONS

TOP SPEED:	155km/h (96mph)
0–96KM/H (0–60MPH):	12.3secs
ENGINE TYPE:	in-line four
DISPLACEMENT:	1275cc (78ci)
WEIGHT:	868kg (1929lb)
MILEAGE:	8.1l/100km (35mpg)

Left: *The Miracle Maestro – Driving is Believing. Until you've had a spell behind the wheel, it's hard to believe how bad the Maestro actually is ...*

Styling wasn't the Maestro's strong point: its scalloped sides and bulbous rear end sat awkwardly with the excessive glass area, and sleeker rivals made the car look dated at launch. Regardless, it was to remain in production for 12 years.

You could choose from either the awful R-Series 1.6-litre (98ci) engine, the dated 1.3-litre (79ci) A-Series or a distinctly agricultural-sounding diesel. None was great, but at least the gearbox was fairly pleasant. It was borrowed from the Volkswagen Golf …

Boy, could the Maestro rust: the paint finish was awful, especially for metallics, and it rotted from the inside out. Rear wheel arches and doors were common rust spots, while doors occasionally drooped due to metal fatigue in the hinges.

You could tell which Maestros were posh simply by looking. Upmarket models got plastic moulded fenders in the same colour as the bodywork. More basic versions came with black-painted metal ones.

AUSTIN MAXI (1969–81)

As an idea, there was nothing wrong with the Austin Maxi. British Leyland wanted a five-door hatchback to compete with some of Europe's top models, by offering comfort for a family of five and becoming one of the most practical cars on the market. All very honourable intentions, but the Maxi's biggest problem was in its execution. Not only was it assembled with all the precision of a 10-year old playing with a Meccano set, but it was also generally unpleasant to drive.

The over-tall overhead cam engines lacked power and were prone to cam chain failure, while the gearbox was initially operated by a series of elastic bands, which gave an awful shift action and were prone to failure. Later cars had a mechanical transmission, but this didn't prove much better …

SPECIFICATIONS

TOP SPEED:	144km/h (89mph)
0–96KM/H (0–60MPH):	15.8secs
ENGINE TYPE:	in-line four
DISPLACEMENT:	1748cc (107ci)
WEIGHT:	972kg (2160lb)
MILEAGE:	9.4l/100km (30mpg)

Left: ***What goes on top of most cars goes inside a Maxi – which is why the footwells were always full of rainwater!***

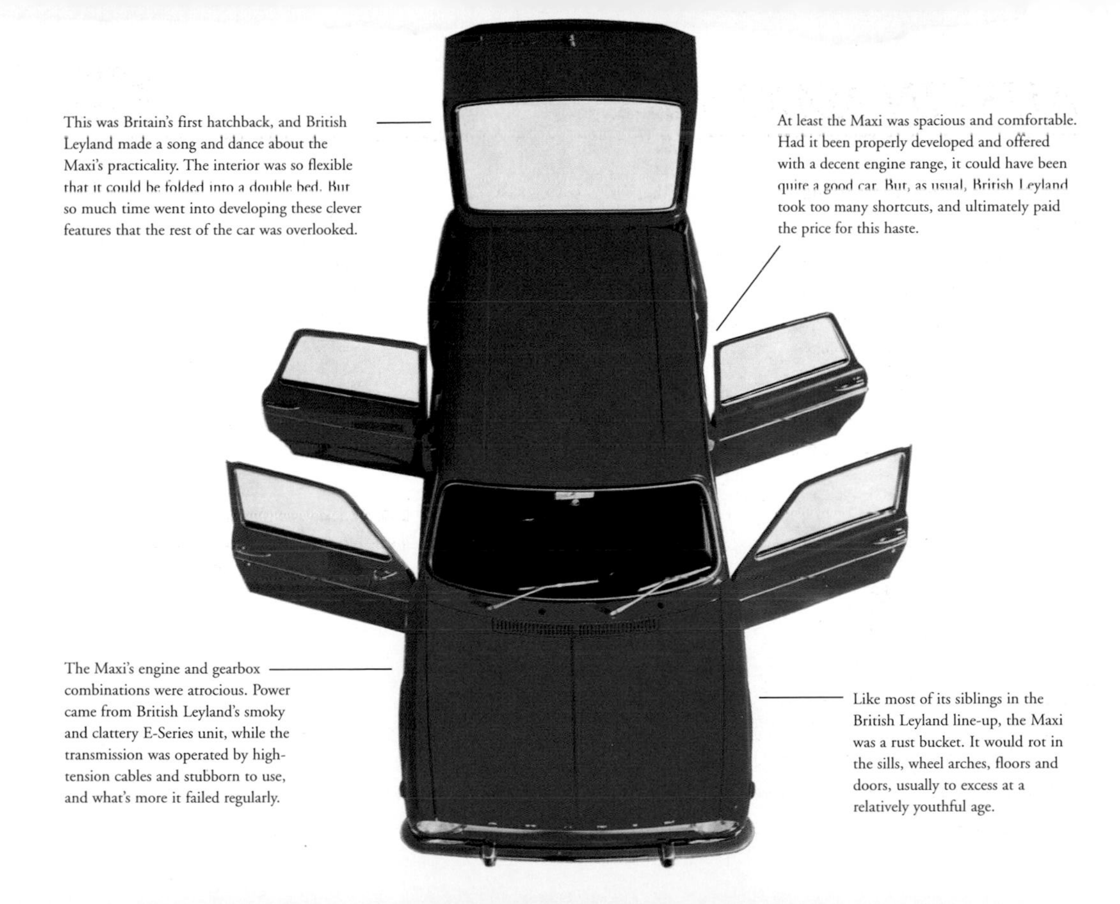

This was Britain's first hatchback, and British Leyland made a song and dance about the Maxi's practicality. The interior was so flexible that it could be folded into a double bed. But so much time went into developing these clever features that the rest of the car was overlooked.

At least the Maxi was spacious and comfortable. Had it been properly developed and offered with a decent engine range, it could have been quite a good car. But, as usual, British Leyland took too many shortcuts, and ultimately paid the price for this haste.

The Maxi's engine and gearbox combinations were atrocious. Power came from British Leyland's smoky and clattery E-Series unit, while the transmission was operated by high-tension cables and stubborn to use, and what's more it failed regularly.

Like most of its siblings in the British Leyland line-up, the Maxi was a rust bucket. It would rot in the sills, wheel arches, floors and doors, usually to excess at a relatively youthful age.

CHEVROLET CAPRICE (1976–90)

If you were looking for excitement in the mid-1970s, the Chevrolet Caprice was not the place to find it. A clumsy mover that wouldn't go round corners at more than walking pace, the Caprice was a real shocker in terms of its build. Almost every model had nasty plastic trim and rampant corrosion; Station Wagon models also got garish fake wood panelling on the sides. Yet the Caprice stayed in production for years. Later models had V8 engines and were restyled in the late 1980s to become Caprice Classic. But even these faster cars understeered and sent reverberations from every bump or pothole from the suspension straight to your clenched-in-anticipation buttocks. It was also one of the most boring-looking cars ever designed.

SPECIFICATIONS

TOP SPEED:	162km/h (100mph)
0–96KM/H (0–60MPH):	13.9secs
ENGINE TYPE:	V8
DISPLACEMENT:	5002cc (305ci)
WEIGHT:	1829kg (4066lb)
MILEAGE:	19.7l/100km (15mpg)

Left: *By far the best part of the Caprice was its engine – it used GM's tried and trusted small-block V8, which is a design classic. It's a shame the rest of the car was so dull.*

The Chevy's one saving grace should have been its engine. A proven General Motors V8 unit, it was refined and fairly lively. But emissions legislation soon put a stop to that, and by the time it reached production it was detuned to be frustratingly slow.

Floors went rusty, often in random patches where the underseal had been applied too thinly. The same was true for the panels, with holes quickly appearing in the trunk lid, doors and front wings.

Inside, the Caprice was even less inspiring than it looked from the outside. The dashboard was made of shiny plastic, often finished in shoe-polish brown, while top models got self-adhesive fake wood inserts.

The Caprice was dreadful to drive: the steering was inaccurate, the ride was far too soft and bouncy and you'd induce tyre squeal if you entered a bend at anything more than walking pace.

CHEVROLET CITATION (1980–86)

With modern construction and styling, plus the added attraction of a hatchback, the Citation was bound to be a success for Chevrolet. And indeed it was, but selling in reasonable numbers doesn't make a car good, as many a former owner will testify. Among the Citation's more unusual idiosyncracies were brakes that stopped working without much in the way of warning, and wear to the steering bushes that made negotiating bends a hit-or-miss affair, in the most literal sense.

Such was early demand for the model that GM let quality slip to keep up with demand, and many cars left the factory with large chunks of underseal missing, leading to excessive corrosion at a very early age.

Left: ***They might have been 'the people who know what performance is all about', but the makers obviously didn't bother applying it to the Citation's dynamics.***

SPECIFICATIONS

TOP SPEED:	193km/h (120mph)
0–96KM/H (0–60MPH):	9.2secs
ENGINE TYPE:	V6
DISPLACEMENT:	2741cc (168ci)
WEIGHT:	1080kg (2400lb)
MILEAGE:	15.7l/100km (18mpg)

Oh, dear! GM was in such a rush to get the Citation to showrooms that quality control slipped. It soon acquired a reputation as one of America's biggest lemons, thanks to its tendency to rust.

Brake failure was rife. The hoses were fitted too tightly and the amount of flexing in the hydraulic system occasionally caused them to work loose, dramatically reducing stopping power.

Handling was not the Citation's strong point. It was prone to oversteer if you lifted off the throttle in bends, while the soft suspension and over-light steering made it difficult to attack corners with any real accuracy.

Citations received a fuel-efficient, six-cylinder engine. But it was a horribly dull unit, with hardly any performance and an excessive thirst for engine oil.

CHEVROLET NOVA *(1970–79)*

The Nova could have been, and should have been, a very good car. Its styling was in keeping with the tastes of its era, and the design was spacious and reasonably economical. But it was General Motors' cost-cutting that caused damage to the model's reputation. Built as an entry-level six-cylinder model for families on a budget, it was sparsely equipped, while the cabin was trimmed in wall-to-wall black plastic. Spending time inside one was like spending time in purgatory thanks to brittle plastics that snapped easily, while excessive oil consumption and rust around the rear wheel arches did nothing to enhance the model's reputation.

Owners also reported regular gearbox faults and cylinder head problems, while some models suffered from inexplicably rotten floorpans. The SS396 V8-engine sports model, on the other hand, was really quite interesting …

Left: ***If you were prepared to 'take it to the limit', you needed nerves of steel. Despite the advertising campaign, the Nova was not a focused driver's car.***

SPECIFICATIONS

TOP SPEED:	153km/h (95mph)
0–96KM/H (0–60MPH):	16.4secs
ENGINE TYPE:	in-line four
DISPLACEMENT:	2286cc (139ci)
WEIGHT:	1010kg (2246lb)
MILEAGE:	11.8l/100km (24mpg)

It looked okay from the outside, but once you climbed inside a Nova, you knew you were in a cheap car. The plastics were shiny and brittle, while the black PVC seats got so hot that they burned bare skin in sunny weather.

Oddly, the Nova was fairly rust-resistant around its front end, but the rear used to rot with alarming speed. It made the car look like it was in fact two different models welded together, as the front was usually in far better condition!

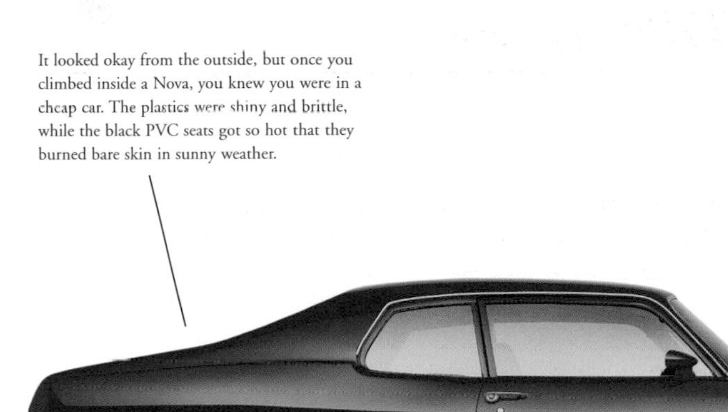

The Nova was very much a nonentity in the handling department. There were cars that cornered better than it did, but it wasn't as bad as many softly sprung and unpredictable American sedans of the 1970s. At least, not until the suspension bushes wore out, which they often did.

With six cylinders and reasonable fuel economy, the Nova's engine looked attractive to buyers who were on a budget. But while it was fairly thrifty at the pumps, it guzzled oil like it was going out of fashion.

DATSUN 120Y *(1974–78)*

The 120Y – also known as 'Sunny' – put Japan's fledgling motor industry firmly on the map, racking up an astonishing number of sales: 2.4 million across Europe and America in a four-year production run. It did this by offering value for money and generous specification levels. Despite the model's early promise, owners discovered in later years that the car was far less good a package than they first thought.

While the cars were mechanically almost unbreakable, it soon became apparent that the Japanese weren't very good at rust-proofing, and the 120Y required a reputation on the secondhand market as something of a rot box, with rust eating its way into the sills, trunk floors, firewalls and subframes faster than a plague of woodworm in a timberyard.

SPECIFICATIONS

TOP SPEED:	145km/h (90mph)
0–96KM/H (0–60MPH):	16.0secs
ENGINE TYPE:	in-line four
DISPLACEMENT:	1171cc (71ci)
WEIGHT:	791kg (1757lb)
MILEAGE:	8.3l/100km (34mpg)

Left: ***The 120Y was a fairly successful rally car, but its poor handling could catch out even the most skilled of drivers on poor surfaces.***

Rot wasn't restricted to the underneath of the vehicle, and many 120Ys started to look shabby very early on. The front panel used to rot so badly that the headlights would sometimes work loose and fall out.

It looked attractive in the brochure. The Datsun had a radio, a heater, reclining seats and a heated rear window as standard – but, though good value, it felt cheap and was poorly finished.

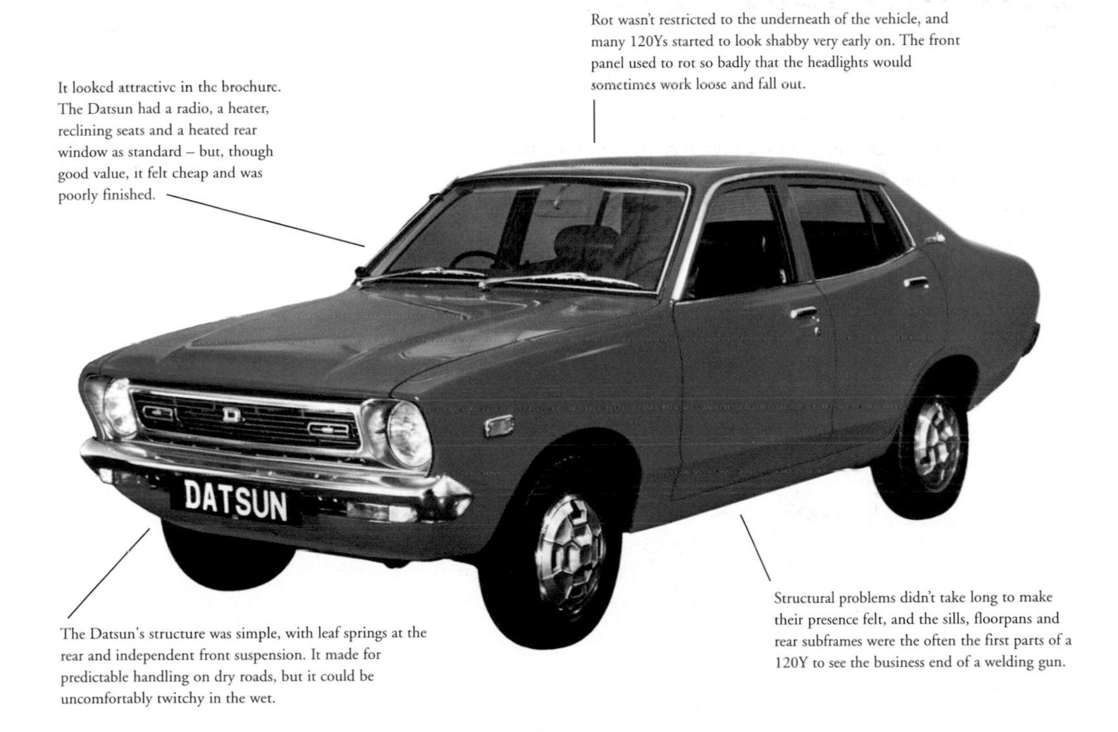

The Datsun's structure was simple, with leaf springs at the rear and independent front suspension. It made for predictable handling on dry roads, but it could be uncomfortably twitchy in the wet.

Structural problems didn't take long to make their presence felt, and the sills, floorpans and rear subframes were the often the first parts of a 120Y to see the business end of a welding gun.

DODGE DART *(1963–66)*

Initially a pared-down version of the Plymouth Valiant, the dodge Dart became a model in its own right in 1963. It was Dodge's first entry into the compact market, and almost every critic agreed that the car's styling simply didn't work, with a stubby trunk and a strange radiator grille and headlight arrangement. But that wasn't the Dart's only problem. Like many early unitary construction models, it was also prone to corrosion and the slant-six engine wasn't renowned for its reliability – although a 'compact' by American terms, the Dart was still a heavy car.

And if the engine was starved of oil, this put the Dodge under pressure, leading to premature bore and cylinder wear. It certainly wasn't one of the company's greatest successes.

SPECIFICATIONS

TOP SPEED:	162km/h (100mph)
0–96KM/H (0–60MPH):	13.9secs
ENGINE TYPE:	slant-six
DISPLACEMENT:	2784cc (170ci)
WEIGHT:	1185kg (2634lb)
MILEAGE:	14.1l/100km (20mpg)

Left: ***'Nothing's going to hold you back!' said Dodge, ignoring the Dart's untidy styling and lame performance.***

No matter what angle you looked at it from, the Dodge Dart was an uncomfortable mishmash of conflicting angles and awkward curves. Unfortunate, given the American market's demand for stylish lines.

The Dart was meant to be an economy car, which meant that many buyers were prepared to accept distinctly average performance. This normally meant better fuel economy, but the Dart was so heavy that the benefit was lost.

Although it resisted rust better than many of its rivals, the Dart was notoriously weak around the rear spring mounts. There was a water trap above the rear subframe, and it was often difficult to see the true extent of corrosion here.

With soggy springs, a long wheelbase and over-assisted power steering, the Dart was a handful to drive, with unpredictable handling and absolutely no sporting appeal.

FIAT CROMA (1986–95)

Fiat has a long-standing reputation for making some of the finest small cars in the world – and quite rightly so – but when it comes to big saloons, the Italian maker has never excelled. The Croma came as part of an Alliance with Saab, Fiat, Lancia and Alfa Romeo, and its platform also underpinned the Saab 9000, Lancia Thema and Alfa Romeo's 164. But of the four, the Croma was by far the least spectacular, with hideously bland styling and a cabin trimmed in unpleasant, low-rent plastics and tweed seat facings.

Rusty doors were always a problem for Croma, but not as much as its electrical problems, which plagued it from early in its life until its death in 1995. Faulty lights, dead heaters and engine management failures were rife.

SPECIFICATIONS

TOP SPEED:	179km/h (111mph)
0–96KM/H (0–60MPH):	10.8secs
ENGINE TYPE:	in-line four
DISPLACEMENT:	1995cc (122ci)
WEIGHT:	1075kg (2390lb)
MILEAGE:	10.4l/100km (27mpg)

Left: *This cutaway shows exactly how conventional the Croma was: it offered no new design ideas. Still, it was fairly spacious.*

Electrical faults were rife, and the dashboard gauges could never be trusted, giving random readouts and indicating faults that didn't exist. Electric windows often packed up, while brake lights and indicators also gave up the ghost.

Like many Italian cars, the Croma's cabin layout was illogical, and some of the switches were in ridiculous locations. To make matters worse, the driving position dictated that you needed short legs and extraordinarily long arms – otherwise it was impossible to sit comfortably.

If Fiat were to succeed in the hotly contested large-car market, it needed a car that had a large dose of Italian styling panache. The Croma wasn't it – its looks were tedious in the extreme and utterly out of keeping with the car's proposed executive image.

Rust was rarely of structural concern, but the panels were quick to spoil, and the doors, hood and tailgate often burst out into random patterns of bubbles.

FIAT STRADA *(1979–88)*

Built by robots – driven by idiots! That was the slogan unfairly applied to the Strada by stand-up comedians after the Italian company tried to play up the model's entirely automated production line at its 1979 launch. The move quickly backfired: using an entire staff of robots led to some horrific issues of build quality. Not only was the Strada plagued by rampant rust problems, as the robots obviously weren't too clever with the Waxoyl, but also it suffered from a raft of electrical failures and, bizarrely, a desire to munch its way through exhaust silencers quicker than any other car on sale. The Strada was sold as the Ritmo in many European markets, and in Spain it was offered as the SEAT Ronda.

The Strada. Handbuilt by robots. The robots. Handbuilt by Fiat.

The new Strada range from £3,845.

FIAT

Fiat Auto. The best selling cars in Europe.

SPECIFICATIONS

TOP SPEED:	140km/h (87mph)
0–96KM/H (0–60MPH):	15.6secs
ENGINE TYPE:	in-line four
DISPLACEMENT:	1301cc (79ci)
WEIGHT:	900kg (2000lb)
MILEAGE:	8.5l/100km (33mpg)

Left: ***As if advertising that the Strada was 'handbuilt by robots' weren't bad enough, suggesting that the robots were 'handbuilt by Fiat' was surely enough to put most prospective buyers off.***

The robots weren't too good in the styling department, either, if the Strada's goofy looks were anything to go by. The car looked like an Austin Maestro that had been pinched at both ends.

The outside was certainly bad, but when it came to the interior the Strada was equally horrible, with a chunky plastic instrument binnacle and thin fabric seat facings. They were usually damp inside, too, as the door seals often leaked.

Rust, rust and more rust: the Strada's fully robotized production line made no provision for ensuring that cars were properly protected when they left the line, and the quality of the metal was so poor that the Strada was renowned for being structurally very weak.

Despite its many problems, the Strada was quite satisfying to drive. Even the smaller engines were fairly lively, while the handling was entertaining. There was even a sporty 130TC version, which was more than a match for the VW Golf GTI.

FORD MAVERICK *(1969–73)*

Yet another car that seemed fine when it was new, but which turned out to be a badly built disaster on wheels, the Ford Maverick lost its reputation to metal fatigue. There were four-door and two-door models, as well as a Mercury version called the Comet. Most were finished in lurid colours with vinyl roofs, and what they all had in common was that they were built on a platform which was heavily prone to rusting.

Nor was their case helped by gutless and fragile in-line six engines. But the cars were cheap and therefore popular, especially during the 1970s fuel crises. While the perilous handling and poor performance were accepted by buyers in the first place, the rot that manifested itself later most certainly wasn't …

SPECIFICATIONS

TOP SPEED:	162km/h (100mph)
0–96KM/H (0–60MPH):	15.4secs
ENGINE TYPE:	in-line six
DISPLACEMENT:	2784cc (170ci)
WEIGHT:	1084kg (2411lb)
MILEAGE:	12.8l/100km (22mpg)

Left: ***The Maverick had many problems, the biggest being that, whichever angle you viewed it from, it was desperately ugly.***

It was spacious and comfortable, but the Maverick offered few creature comforts. The cabin was normally trimmed in cheap-looking black vinyl, which got unbearably hot during summer months.
Not even the styling could save the Maverick. It wasn't especially offensive, but it was considered rather bland, while two-door versions had far too much metal in the rear quarter panel – a problem, given how rust-prone the metal was.
1972
Ford dropped a blooper with the original Maverick. The floorpan had so many dirt and water traps that it was inevitable rust would eventually take hold, particularly as the cars left the factory with hardly any underseal.
The car's soft ride and cart-sprung suspension meant handling wasn't the Maverick's forte, while the steering was vague, especially when the front bushes and gaiters started to wear out, causing the mechanism to seize.

FORD ZEPHYR MK 4 (1966–72)

Sixteen years after the first Zephyr arrived, the final incarnation of Ford's executive saloon came on the scene. In four generations, it had grown considerably. The Zephyr had always echoed the styling of American models, starting off as a rounded, shapely replica of a 1950s sedan, and culminating in this – a gargantuan, incongruous slab of faux-Detroit iron.

Critics quickly jumped on the Zephyr's awkward stubby-tailed, long-nosed styling, likening its front end to the landing deck of an aircraft carrier. Its metal was typical of the 1960s and that meant it had rust traps in abundance. The Zephyr may have been comfortable, and in its upmarket Executive form it was really quite a plush car, but it was built to a price – and it showed.

SPECIFICATIONS

TOP SPEED:	166km/h (103mph)
0–96KM/H (0–60MPH):	13.4secs
ENGINE TYPE:	V6
DISPLACEMENT:	2495cc (151ci)
WEIGHT:	1310kg (2912lb)
MILEAGE:	13.4l/100km (21mpg)

Left: ***There were acres of space inside the Zephyr Mk 4, but low-spec models such as this example were not at all luxurious. Check out the black plastic instrument binnacle and nasty brown steering wheel!***

Nobody could pretend that the Zephyr Mk 4 was a pretty car. Its lines were out of proportion, and the fact that the long nose was unnecessary, given that it housed a compact V6 engine, made it even more bizarre.

You'd expect a car this big to have plenty of room for passengers, but the middle section was actually fairly small. Legroom in the back was cramped, forcing Ford to build a stretched 'Executive' model for chauffeur use.

Quite why the trunk was so stubby, nobody knows – it certainly didn't do the styling any favours. It was so cramped inside that Ford had to move the spare wheel to under the hood, where there was plenty of unused space.

The Mk 4 Zephyr was never going to hurtle round racetracks, but buyers were stalled by its reluctance to go round any corners at all. The steering was utterly lifeless, while the skinny tyres had their work cut out trying to convince the huge mass to change direction.

HYUNDAI STELLAR (1984–91)

Korea's answer to the Ford Cortina was exactly that – a car that more or less emulated Ford's family favourite in every way. With boxy styling, MacPherson struts at the front and a live rear axle on coil springs, the Stellar wasn't bound for stardom. Indeed, it was totally anonymous in almost every respect, while its bouncy ride and unpredictable handling did it few favours. Initially, it was fairly popular among Europe's middle classes, as it offered plenty of space and comfort for not much outlay. But then the rear spring mounts started to rot out, the doors crumbled into piles of iron oxide and the Mitsubishi-copy gearboxes jumped out of reverse, and buyers realized that they'd bought a car that was less attractive than it first seemed.

SPECIFICATIONS

TOP SPEED:	160km/h (98mph)
0–96KM/H (0–60MPH):	14.7secs
ENGINE TYPE:	in-line four
DISPLACEMENT:	1597cc (97ci)
WEIGHT:	1003kg (2231lb)
MILEAGE:	9.7l/100km (29mpg)

Left: ***This Hyundai Stellar is parked on the surface of the moon – the driver lost control on a damp roundabout, and ended up spinning there.***

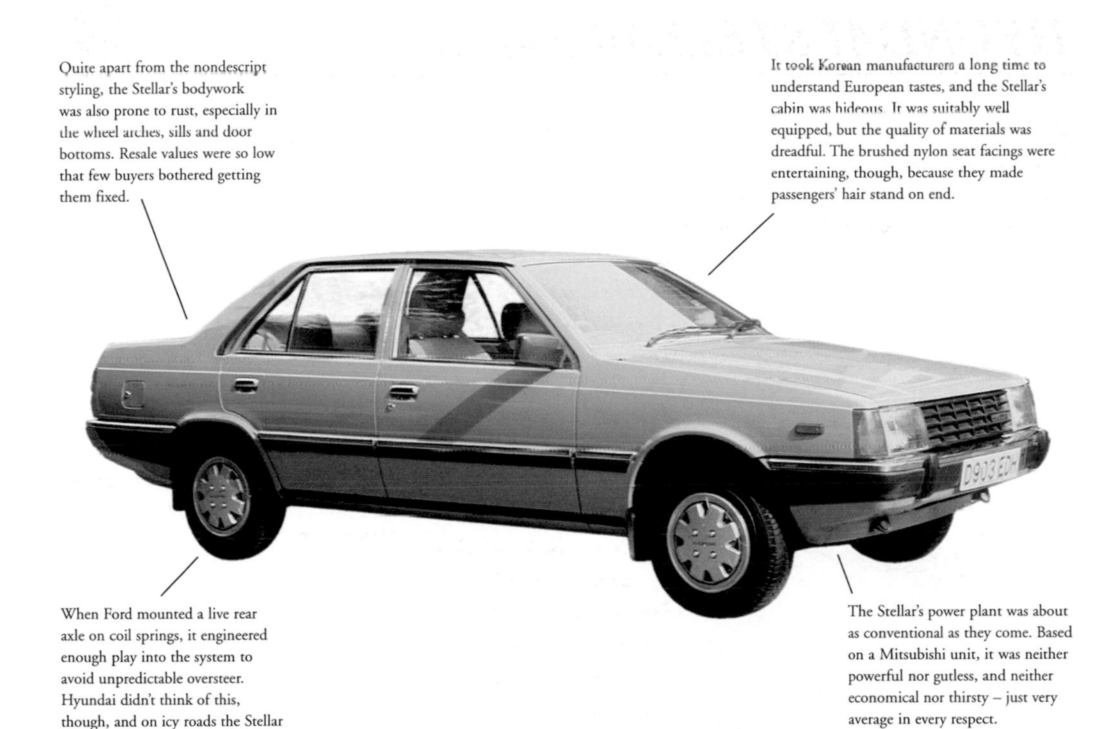

Quite apart from the nondescript styling, the Stellar's bodywork was also prone to rust, especially in the wheel arches, sills and door bottoms. Resale values were so low that few buyers bothered getting them fixed.

It took Korean manufacturers a long time to understand European tastes, and the Stellar's cabin was hideous. It was suitably well equipped, but the quality of materials was dreadful. The brushed nylon seat facings were entertaining, though, because they made passengers' hair stand on end.

When Ford mounted a live rear axle on coil springs, it engineered enough play into the system to avoid unpredictable oversteer. Hyundai didn't think of this, though, and on icy roads the Stellar spun like a drunken figure skater.

The Stellar's power plant was about as conventional as they come. Based on a Mitsubishi unit, it was neither powerful nor gutless, and neither economical nor thirsty – just very average in every respect.

JENSEN HEALEY (1970–76)

Born out of a marriage of convenience, the Jensen Healey was supposed to be the car that rescued Jensen and kept Healey in business, but in reality it turned out to be an out-and-out disaster on a scale peculiarly common to low-volume British manufacturers. Launched in 1972, the two-seater convertible used a Lotus engine, Vauxhall Viva running gear, a Chrysler gearbox and hand-built bodywork, all screwed together at Jensen's factory in the British West Midlands.

Of all the bits, it was only the proven Viva suspension and steering that gave no trouble. The Lotus engines were prone to overheating and warping their cylinder heads, the Chrysler gearboxes were too weak to cope with the power, and the bodywork turned to powder after its first grit-salted winter. Most fatally, though, the car lacked charisma.

SPECIFICATIONS

TOP SPEED:	200km/h (125mph)
0–96KM/H (0–60MPH):	8.8secs
ENGINE TYPE:	in-line four
DISPLACEMENT:	1973cc (120ci)
WEIGHT:	1053kg (2340lb)
MILEAGE:	24mpg (11.7l/100km)

Left: ***The front end of the Jensen Healey had to sit fairly high to accommodate the Lotus engine, as this cutaway drawing shows.***

With the impressive Jensen and Healey names, you'd have expected their collaboration to be both aggressively styled and beautiful – but it was neither. The front end styling had to be modified to satisfy US safety legislation, so it was higher than originally intended. There was also an estate-car GT model, which was an even less attractive proposition.

Rust was always fairly quick to take hold – a symptom of poor quality control and the use of cheap materials to keep costs at a minimum. The first bits to go were usually the doors, but rot also found its way into the rear wheel arches, floorpans and trunk lid. Rather more worryingly, it also found its way into the suspension mounts.

The Jensen Healey was the perfect solution for Lotus, which was trying to swell its coffers by selling its engines to external makers. But just like the engines in the Elite and Eclat, the car's 2.0-litre (122ci) 16-valve power plant was fragile and prone to overheating, leading to cylinder head problems.

You might expect a sports car from two of the finest names in the business to be brimming with advanced chassis technology, but no. The steering came from Ford, the gearbox was a dated Chrysler unit and the suspension was taken wholesale from the humble Vauxhall Viva.

LADA SAMARA *(1989–96)*

After years of churning out recycled Fiat 124s, the Samara was the first entirely new model from AutoVaz (known globally as Lada) in three decades. And, from a distance, it appeared quite respectable, with fairly modern lines and room to seat four in comfort at a cut price. Sadly, though, it soon came to stand for everything its predecessors had done – cheap, inexpensive transport which often reminded you that shopping on the secondhand market would have been a better idea. Flimsy interiors fell apart within weeks, the gearboxes were truly awful and the metalwork was so thin you could make the panels warp by poking them with your little finger. Later cars had to be fitted with catalytic converters to cope with new European emissions legislation, and this meant fitting electronic engine management systems, which packed up with alarming regularity, leading to the premature death of many a fairly new Lada Samara.

SPECIFICATIONS

TOP SPEED:	137km/h (85mph)
0–96KM/H (0–60MPH):	14.1secs
ENGINE TYPE:	in-line four
DISPLACEMENT:	1100cc (67ci)
WEIGHT:	895kg (1989lb)
MILEAGE:	8.3l/100km (34mpg)

Left: ***According to the ad men, the Lada Samara was 'the new car more people are getting into'. If this was true, why did European imports dry up two years after this brochure was published?***

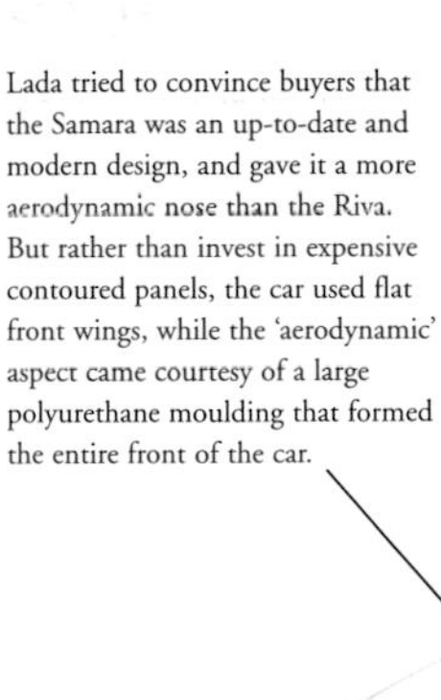

Lada tried to convince buyers that the Samara was an up-to-date and modern design, and gave it a more aerodynamic nose than the Riva. But rather than invest in expensive contoured panels, the car used flat front wings, while the 'aerodynamic' aspect came courtesy of a large polyurethane moulding that formed the entire front of the car.

The Samara's cabin gave no impression of quality. It was made out of plastics that felt extremely cheap, and the one-piece dashboard mouldings were badly made and never quite fitted properly. Also, the chunky switches for the fog lights and heated rear window used to fall off.

The outside was no better built than the inside, with plastic trim that parted company from the rest of the car with surprising ease. Panel gaps varied from one car to another, and it's rumoured that the British importer dismantled and rebuilt every car it brought in, not daring to sell factory-finished models to the public!

The Samara was Lada's first front-wheel-drive car, but it offered very little handling benefit over the old rear-drive models. The crude suspension made for very poor handling, while the driving experience was further hampered by an appalling gear change and an intrusively noisy engine note.

LANCIA DEDRA *(1989–95)*

Lancia tried hard to stave off the reputation for excessive corrosion that had so blighted the company throughout the late 1970s and early 1980s, so it was hard not to feel sorry for the Italian firm when the Dedra arrived on the scene. For here was a car that was rustproofed so well it could have been left simmering in a bath of salt and would still have emerged unscathed. Sadly, though, Lancia spent so much time protecting its cars from rust that it completely overlooked more trivial matters, such as whether or not the vehicles would keep running for as long as the metalwork survived.

And while it is true that the Lancia Dedra never rotted, it all too frequently ground to a halt with any one of a number of electrical maladies. Nor did its clumsy ride and indifferent handling help matters.

SPECIFICATIONS

TOP SPEED:	202km/h (125mph)
0–96KM/H (0–60MPH):	10.0secs
ENGINE TYPE:	in-line four
DISPLACEMENT:	1995cc (122ci)
WEIGHT:	1247kg (2772lb)
MILEAGE:	9.4l/100km (30mpg)

Left: ***Lancia was desperate to shake off its old image and declared that the Dedra was 'built to conquer the toughest roads of Europe'. Sadly, such roads only served to highlight the car's dreadful ride quality.***

One area where Lancia was surprisingly successful with the Dedra was its styling. From the outside, it looked like a modern, upmarket and well-finished family saloon, but owners soon discovered that the elegant exterior concealed some horrors …

Italian cars have never been known for their interior quality, and the Dedra was no exception. The plastics were crudely finished, while buyers often experienced problems with the door-handle mechanisms. They used to snap off inside the door, meaning the only way to get out was to put the window down and use the external handle!

Lancia had a reputation for building cars that had handling flair, so the Dedra's lack of composure won it few fans. The ride used to transmit bumps into the cabin and, although it gripped quite well, the steering offered little in the way of feedback while the platform offered very little in the way of driver involvement.

Reliability wasn't great. The engine itself was fundamentally quite tough, but it was the electrical system that caused all manner of problems. The engine management system was temperamental, causing the engine to lose its fuel supply, while heater fans, electric windows and lights were often prone to failure.

LOTUS ELITE/ECLAT (1974–91)

Lotus thought it was being avant-garde when it unveiled the wedge-shaped Elite and Eclat models in 1974, but in reality it produced a pair of plastic-bodied monstrosities, which embodied all that was gaudy and uncultured about 1970s fashion. Perhaps the cars' only saving grace was their gutsy and responsive 2.2-litre engines, but a lack of development soon put paid to this, with coolant leaks causing them to overheat and warp their head gaskets.

Electrical faults were even more common, usually caused by rusty earth terminals. It's common to see them driving round with one pop-up headlight popped up, and the other one 'sleeping', as a lack of lubrication in the mechanism caused the electric motors inside to burn out. Ironically, as the cheapest entry into Lotus ownership, they have since acquired quite a cult following.

SPECIFICATIONS

TOP SPEED:	200km/h (124mph)
0–96KM/H (0–60MPH):	7.8secs
ENGINE TYPE:	in-line four
DISPLACEMENT:	1973cc (120ci)
WEIGHT:	1126kg (2503lb)
MILEAGE:	10l/100km (28mpg)

Left: *A breath of fresh air from Lotus – or a sharp intake of breath from your local mechanic when he sees that the cylinder head has warped ...*

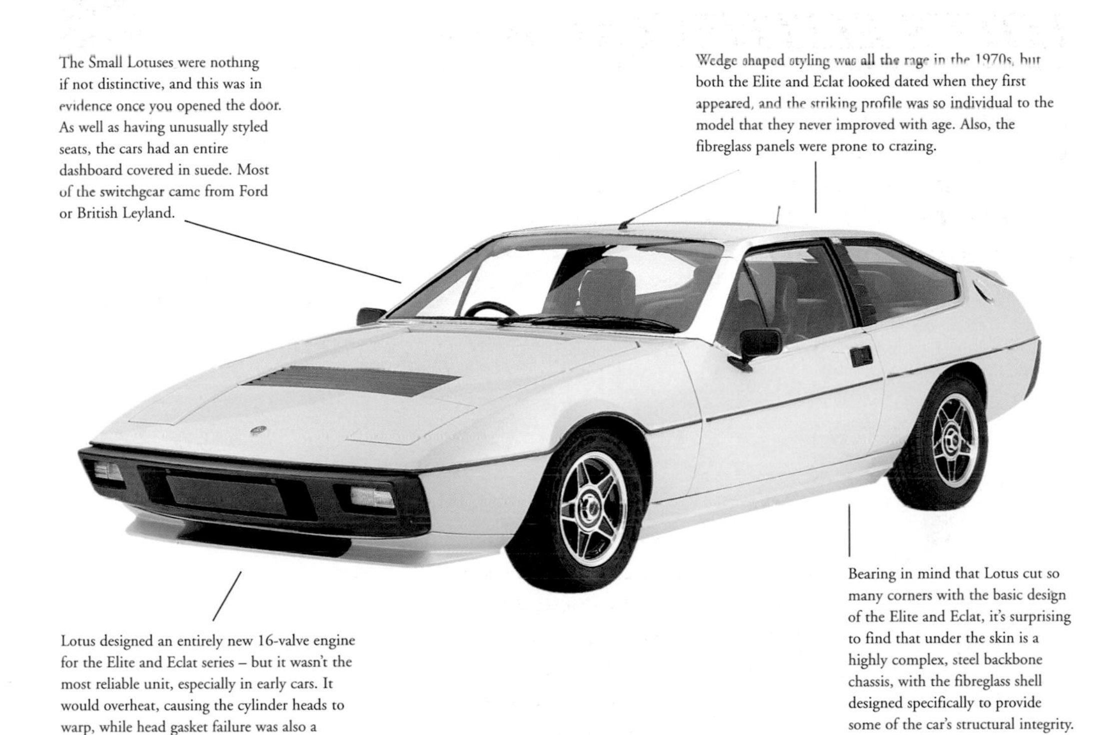

The Small Lotuses were nothing if not distinctive, and this was in evidence once you opened the door. As well as having unusually styled seats, the cars had an entire dashboard covered in suede. Most of the switchgear came from Ford or British Leyland.

Wedge shaped styling was all the rage in the 1970s, but both the Elite and Eclat looked dated when they first appeared, and the striking profile was so individual to the model that they never improved with age. Also, the fibreglass panels were prone to crazing.

Lotus designed an entirely new 16-valve engine for the Elite and Eclat series – but it wasn't the most reliable unit, especially in early cars. It would overheat, causing the cylinder heads to warp, while head gasket failure was also a regular (and expensive) occurrence.

Bearing in mind that Lotus cut so many corners with the basic design of the Elite and Eclat, it's surprising to find that under the skin is a highly complex, steel backbone chassis, with the fibreglass shell designed specifically to provide some of the car's structural integrity.

MAZDA 626 MONTROSE (1978–82)

Naming a car after a granite-clad Scottish backwater was never going to do it any favours, but the Mazda 626 Montrose's problems went far deeper than the badge on its tail. Styled as Japan's answer to the ultra-popular Ford Capri, the Montrose was supposed to be a sporty coupé. But in reality, it was little more than a two-door saloon on a standard 626 platform – a chassis that had never been renowned for its dynamic brilliance.

Importing spares from Japan was astronomically expensive, although this rarely became a problem because the mechanical components usually outlived the cars' rather fragile bodywork, such was the Montrose's endemic rust problem. But Mazda wasn't deterred, and its saloon cars have since become a byword for reliability and build quality.

SPECIFICATIONS

TOP SPEED:	171km/h (106mph)
0–96KM/H (0–60MPH):	11.7secs
ENGINE TYPE:	in-line four
DISPLACEMENT:	1970cc (120ci)
WEIGHT:	1075kg (2389lb)
MILEAGE:	10.8l/100km (26mpg)

Left: ***If the Montrose was supposed to be sporty, the designers obviously forgot the interior. It's a mess of black vinyl and brown velour, much like any other 1970s saloon.***

As Japan's answer to the Ford Cortina, it's no surprise to learn that the Mazda's underpinnings were almost a carbon copy of those of the Ford. The rear axle was a solid beam, mounted on coil springs, which made handling 'interesting' in the wet …

Equipment levels were generous, but the Montrose lacked finesse. The shiny plastics felt very cheap to the touch, and the brittle trim used to rattle. The dashboard was also deeply uninspiring to look at, especially when finished in Japan's trademark shoe-polish brown.

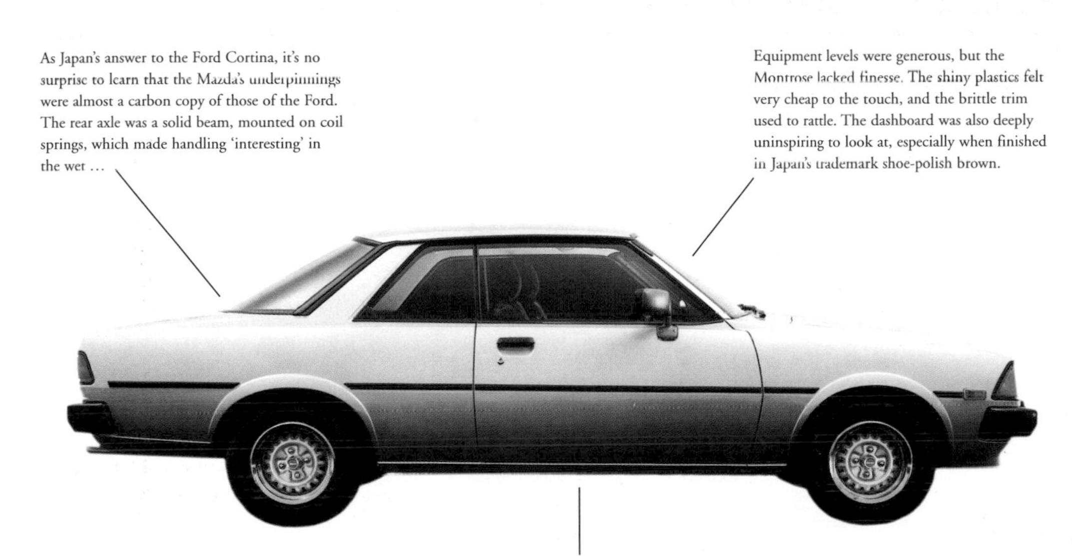

Not only was the Montrose lacking any kind of style or design detailing, but it was also not built for the long term. It came from an era when Japanese car makers had yet to discover the benefits of careful rustproofing, so it suffered from extensive corrosion in the floors, suspension mounts and door bottoms.

MORRIS MARINA (1970–81)

It's difficult to explain the true horror of the Morris Marina, such were its many shortcomings. With British Leyland losing sales to arch rival Ford hand over fist, it had to develop a replacement for the ageing 'Farina' saloon models quickly – the Marina was it. Based on the ancient underpinnings of the Morris Minor, it looked reasonably modern – but that was its only saving grace. At the launch, journalists advised BL to modify the car's suspension, as it was dangerous to drive. BL didn't, and the result was a car that would plough on in a straight line at speed, regardless of steering input. But that was only one of a catalogue of problems, including leaky interiors, rust and collapsing front suspension units.

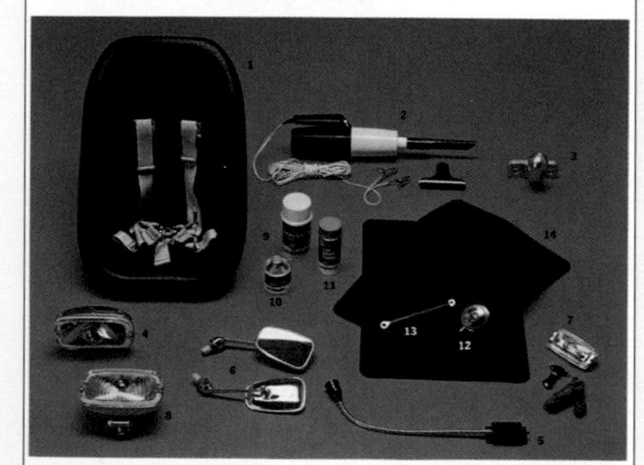

Left: *The Marina wasn't so much 'Beauty with brains behind it' as deathly dull. The accessories were almost laughable compared with what Japanese cars were offering at that time.*

SPECIFICATIONS

TOP SPEED:	161km/h (100mph)
0–96KM/H (0–60MPH):	12.3secs
ENGINE TYPE:	in-line four
DISPLACEMENT:	1798cc (110ci)
WEIGHT:	1150kg (2557lb)
MILEAGE:	10.4l/100km (27mpg)

Convention was the key to the Marina's success, so it looked entirely ordinary in saloon or estate form. The coupé, on the other hand, was quite interestingly styled, with a swooping roofline and stubby rear. But whichever body style you chose, you could expect leaky window seals, rot in the front wings and A-pillars, and crumbling sills.

Later models got a less than successful facelift, which did away with the original car's neat chrome fenders and aluminium grille. They got black painted fenders, a plastic grille and a foam front spoiler, which ruined the purity of the car's lines, especially on the coupé. The handling was much improved, though!

The Marina was based on the Morris Minor – an ancient design. It showed in the car's handling, which was at best wayward and at worst downright terrifying. Larger-engined models were by far the worst, as the extra weight over the front end combined with the antiquated lever-arm suspension made for drastic understeer.

The Marina's brakes were initially so bad that road testers at the launch told British Leyland to correct the problem immediately, or they'd report them as dangerous in the press. British Leyland told the critics that the problem had been resolved, so most of the initial reviews were quite polite. Then journalists got to drive further examples and discovered that, despite its claims, the company hadn't improved the brakes at all …

PEUGEOT 604 (1975–86)

Few people remember the 604, and that's hardly surprising given its styling, which is so dull it could have been the template for a three-box saloon. Still, the 604 was truly enormous and extremely comfortable, despite its lack of styling drama. Where it fell down was in the execution. French executive cars have never had a great reputation for build quality, and the 604 was no exception, with trim that worked itself free of the bodywork and sills that rotted from the inside out, so you couldn't see the rust coming through until it was too late to do anything about it.

It was also unrefined, and drank fuel in V6 gasoline guise. In its defence, it was one of the first commercially successful diesel-powered cars, and the 604 provided stalwart service to many a Parisian taxi driver with a 2.3-litre oil-burner under the hood.

SPECIFICATIONS

TOP SPEED:	190km/h (118mph)
0–96KM/H (0–60MPH):	9.4secs
ENGINE TYPE:	V6
DISPLACEMENT:	2849cc (173ci)
WEIGHT:	1408kg (3131lb)
MILEAGE:	12.8l/100km (22mpg)

Left: *It might have been 'The best Peugeot in the world' when it came out, but the 604 was a long way from being one of the best cars in the world.*

The 604 didn't have panels so much as enormous slabs of metal, and the uncompromising styling didn't lend itself to rust prevention. The doors were usually the first to go, followed by the rear valance and the front scuttle panel.

While it may not have been a great car, at least the 604 was comfortable. Exceptionally comfortable, in fact, with rear legroom to rival that of a Rolls-Royce Silver Shadow.

Peugeot wanted to make sure the 604 had class-leading ride, so the suspension was deliberately made to be very soft to aid shock absorption. But, in reality, this just made it bouncy, while the handling was indifferent, to say the least.

Power came from a choice of V6 gasoline engines, or a 2.3-litre (140ci) diesel. The derv-powered unit was a first in the executive car class, but it was dreadfully slow and lacking in refinement. Gasoline units were much better, but these had an alarming (and expensive) thirst for fuel.

RENAULT 14 *(1977–83)*

Earning the dubious title of 'France's rustiest car', the Renault 14 was all too quick to burst into a dose of orangey-brown measles at the first sign of a shower. Not only that, but the wiring looms used to work free of the firewalls as well, causing all manner of electrical failures and occasionally the odd fire, if the wrong two wires touched. Not an auspicious start, then.

What's more, the Renault 14 was a truly hideous piece of styling, complete with odd pear-shaped flanks, and was also unbelievably dull to drive. It's hard to find anything interesting to say about it, apart from the fact that the wheelbase was slightly longer on one side than the other, due to the angle at which the drive shafts emerged from the engine. Strange, but true …

SPECIFICATIONS

TOP SPEED:	138km/h (89mph)
0–96KM/H (0–60MPH):	15.3secs
ENGINE TYPE:	in-line four
DISPLACEMENT:	1218cc (74ci)
WEIGHT:	858kg (1906lb)
MILEAGE:	7.8l/100km (36mpg)

Left: *Putting this amount of luggage in the trunk of a Renault 14 was a brave move. So endemic was the car's rust problem that the trunk floor simply fell through if the load was too heavy.*

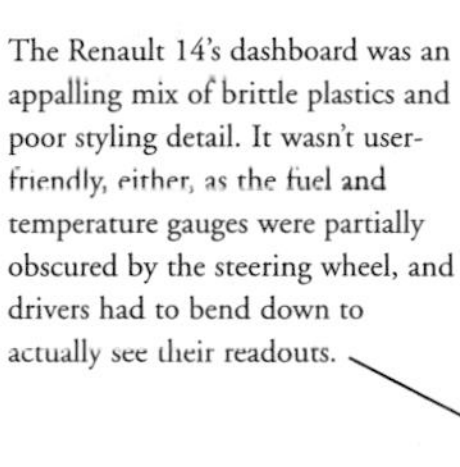

The Renault 14's dashboard was an appalling mix of brittle plastics and poor styling detail. It wasn't user-friendly, either, as the fuel and temperature gauges were partially obscured by the steering wheel, and drivers had to bend down to actually see their readouts.

Rust was a major problem. The 14 had an Alfasud-like ability to rot, with water traps between the rear door shuts and wheel arches causing the rear of the sills to rot with surprising vigour. The trunk floor, inner wings, firewall and front valance were also extraordinarily rust-prone.

Renault used some very unusual packaging under the hood – the engine was a little too wide to be mounted horizontally, so it was put in at an angle, with drive shafts jutting out from different places on each side. The gearbox was in the sump and was lubricated by the engine oil.

If the 14 could be praised in any area, then it was the handling and ride. These weren't sensational, but at least it had reasonable grip and offered decent levels of comfort. But this was hardly a redeeming feature, given the car's all-round awfulness!

RENAULT FUEGO (1981–85)

It was Renault's first car in over 20 years to be given a name and not a number, but that didn't make the Fuego an especially desirable model. Indeed, its name was rather unfortunate. 'Fuego' is Spanish for fire, and the Renault 18-based coupé had a rather worrying habit of bursting into flames without warning because its electrical circuitry was badly laid out and prone to shorting. Then there were the car's engines – with a choice of antiquated 1.4-litre or 1.7-litre overhead valve four-pots under the hood, the Fuego was certainly no fire-breather.

Renault tried to counter this by adding a 2.0-litre turbo, at which point they discovered it also had dreadful handling. Undeterred, the French maker tried to foist the Fuego on the American market as an AMC, where it was an even more spectacular failure. Still, it was quite pretty …

SPECIFICATIONS

TOP SPEED:	183km/h (114mph)
0–96KM/H (0–60MPH):	9.4secs
ENGINE TYPE:	in-line four
DISPLACEMENT:	1995cc (122ci)
WEIGHT:	1042kg (2317lb)
MILEAGE:	10.8l/100km (26mpg)

Left: ***Renault took an unusual approach with the Fuego brochure – the cover had acres of blank space, and only a tiny picture of the car. It's almost as if they were ashamed of it!***

Fire by name, and fire by nature – early Fuegos suffered from electrical faults, which got very nasty when sparks found their way into the fuel system. Other reliability issues included weak gear linkages and premature driveshaft wear on the Turbo.

It might have been billed as a sports car, but the entry-level Fuego was certainly no ball of fire, offering just 64bhp from its 1.4-litre (85ci) engine. The 1.6-litre (98ci) Turbo model offered almost twice the power output, but suffered from turbo lag and immense wheelspin.

When it came to rust resistance, the Fuego was no better than any other Renault of the era. That meant it was prone to corrosion, with the most vulnerable areas being the rear arches and front inner wings.

There was nothing sporting about the Fuego's chassis. Just a standard Renault 18 with slightly firmer dampers – and it showed. The Fuego was prone to understeer, and the livelier models simply couldn't cope with the power.

ROVER SD1 *(1976–1987)*

Depending on which camp you fall into, the Rover SD1 is either one of the greatest cars of all time or one of the most horrific. In design terms, it was certainly something quite special. Styled by Briton David Bache, it took cues from the Ferrari Daytona to give it a purposeful and dramatic appearance, while a glorious V8 engine and assured handling, coupled to great levels of luxury, made it immensely desirable. But it was built and developed by British Leyland, and that meant the SD1 had a difficult birth.

The earliest cars were prone to body rot, and the non-V8 engines were slovenly and prone to failure. Then there were the electrics, which packed up on a regular basis, leaving many unfortunate owners clambering out of the car's sunroof when its central locking tried to imprison them. A brave effort, but ultimately doomed to failure.

SPECIFICATIONS

TOP SPEED:	203km/h (126mph)
0–96KM/H (0–60MPH):	8.4secs
ENGINE TYPE:	V8
DISPLACEMENT:	3528cc (215ci)
WEIGHT:	1345kg (2989lb)
MILEAGE:	12.3l/100km (23mpg)

Left: ***Britain's Car magazine warned the likes of BMW, Volvo, Mercedes and Citroën to beware when the SD1 appeared – but that was before anyone understood how troublesome the newcomer would be.***

One clever aspect of the SD1 was its dashboard. To reduce build costs for both left and right-hand-drive models, the car came with two holes pre drilled in the firewall for the steering column. The instrument pack could then be slid across the fascia, to fit on either side, with all the minor controls mounted in a centre stack.

Rover launched a new range of six-cylinder engines in the SD1, but they weren't great. Performance was poor, fuel economy was unimpressive and reliability was questionable. The lively and dependable V8 was a far better choice, if you could afford the fuel bills.

The first SD1's suffered from shocking build quality, and the paint on metallic models used to flake off, leading to horrific corrosion problems. Later cars proved much better, but the car's reputation never really recovered.

SKODA ESTELLE (1975–91)

The Skoda Estelle was a car built for a purpose, which was to transport the citizens of what was then Czechosolvakia for as little outlay as possible. Little thought was given to the driving experience: putting the engine in the back was considered an effective cost-cutting move, as was installing swing-arm rear suspension, which gave it lethal handling. That said, the dramatic oversteer made it fairly popular with the rallying fraternity!

Inside, the Estelle was trimmed in austere shiny black plastic and had column stalks that snapped off in your hands, while for racier buyers there was the option of the 'Rapid' coupé, a hastily conceived (and shocking to drive) convertible. Still, once the initial build-quality problems had been overcome by its frustrated owner, an Estelle would run for ever – although the engine needed regular fettling.

SPECIFICATIONS

TOP SPEED:	135km/h (84mph)
0–96KM/H (0–60MPH):	18.9secs
ENGINE TYPE:	in-line four
DISPLACEMENT:	1174cc (72ci)
WEIGHT:	873kg (1940lb)
MILEAGE:	8.3l/100km (34mpg)

Left: ***It might not have been a great road car, but rally drivers loved the Skoda's simple mechanical layout and tail-happy handling characteristics.***

It looked like a conventional front-engine saloon, but because the Estelle's engine was in the rear the front was instead a fairly generous luggage bay. Some owners filled this with a bag of cement to even out the car's weight distribution and improve its handling immensely!

Luxuries weren't high on the list of Skoda buyers' priorities, and that showed as soon as you opened the door. The cabin was coated in black PVC, while the effect was made even worse by the incredibly poor quality of the plastics. It wasn't especially comfortable, either.

Despite its bargain-basement price tag, the Estelle was remarkably tough. It resisted rust far better than many of its contemporaries, meaning that it could easily last for years. If only it had not been so dreadful to drive, it could have been quite a good car …

The Estelle had its all-alloy engine mounted behind the rear wheels, and the back suspension came with swing axles as standard. That made it dramatically tail-happy, and the Estelle's wayward handling was always heavily criticized.

TRIUMPH STAG *(1970–77)*

If proof were needed that British Leyland was the most frustrating car company on Earth, then the Stag was it. The designers produced a car that looked stunning, and it was based on the Triumph 2000 platform, meaning that it was also inexpensive to build. This should have freed up some money for proper development, but sadly the accountants forgot to tell the workers.

The Stag, which could easily have been fitted with BL's Rover V8 engine, instead got its own alloy-head V8, effectively created by welding two Dolomite 1500 units together. The result was a reaction between the steel block and alloy head, which led to blown head gaskets, overheating and a warped top end of the engine before most had covered 80,000km (50,000 miles).

Many owners did what BL should have done and put their own Rover V8 engines in – with remarkably successful effects.

SPECIFICATIONS

TOP SPEED:	188km/h (117mph)
0–96KM/H (0–60MPH):	10.1secs
ENGINE TYPE:	V8
DISPLACEMENT:	2997cc (183ci)
WEIGHT:	1263kg (2807lb)
MILEAGE:	11.8l/100km (24mpg)

Left: ***'Keeping up with the Triumph Stag is no problem if you've got the money,' proclaimed British Leyland, almost as if it knew that the new 2+2 cabriolet would hit owners hard in the wallet ...***

The Stag's engine became its downfall. Its unique 3.0-litre (183ci) V8 was expensive to develop, and pointless given the fact that BL already owned the Rover 3.5-litre (214ci) unit. It was badly engineered, and suffered from problems with warped cylinder heads and blown gaskets.

Inside, the Stag was exceptionally comfortable and generously equipped. To keep costs to a minimum, the dashboard and seats were lifted directly from the Triumph 2000, as was the four-speed manual overdrive gearbox.

Like most BL cars, the Stag wasn't immune from the rust bug. Corrosion soon worked its way into the sills, rear quarter panels, trunk lid and front wings, while the firewall edges and fuel tank were also weak areas.

The Stag was based on the platform of the Triumph 2000, and it showed exactly how good that saloon car's platform was. It could have been a brilliant success story, but, as was always the case when British Leyland was involved, it turned into an unreliable, underdeveloped financial nightmare – and a crying shame, to boot …

TRIUMPH TR7 *(1976–83)*

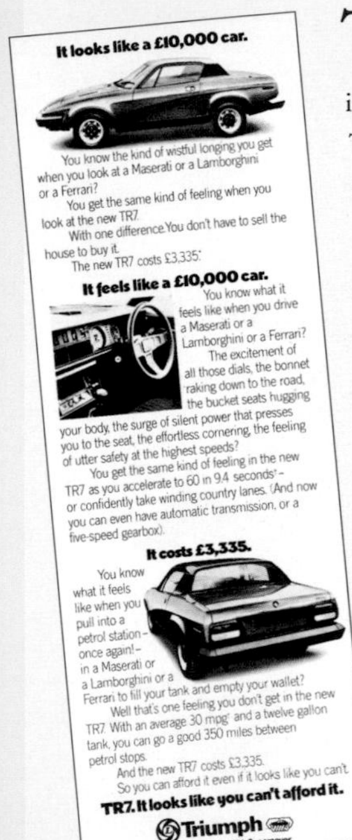

The wedge was king in the 1970s, so British Leyland decided that it should be applied to its replacement for the butch and masculine TR6. The result was the TR7, and when the Italian designer Giorgietto Giuigiaro saw the car for the first time, he walked round it and cried, 'Oh no! They've done the same to the other side as well.'

To cut costs, British Leyland designed the TR7 to house Triumph's excellent 2.0-litre slant-four engine, but the fit was so snug that the proposed 16-valve head from the Dolomite Sprint wouldn't fit under the hood, meaning that it had to make do with an eight-valve top end that could only muster 105bhp. To add insult to injury, the TR7 loved to rust, as many disgruntled owners soon found out.

SPECIFICATIONS

TOP SPEED:	177km/h (110mph)
0–96KM/H (0–60MPH):	9.1secs
ENGINE TYPE:	slant-four
DISPLACEMENT:	1998cc (122ci)
WEIGHT:	992kg (2205lb)
MILEAGE:	10.0l/100km (28mpg)

Left: ***'It looks like you can't afford it.' No, it doesn't. It looks like a cheap imitation of a proper sports car, with the door handles from an Austin Allegro.***

The TR7 was supposed to get the 16-valve cylinder head from the Dolomite Sprint, but nobody told the styling department! When it appeared, the wedge-shaped hood line was too low to shoehorn the tall engine block under, so it had to make do with an eight-valve head instead.

If proof were needed that the 1970s was a decade of questionable taste, then look no further than the interior of a TR7. The leather-clad sport steering wheel was reminiscent of its era, but not as much as the garish tartan-faced seats, which were, frankly, disgusting.

It's no wonder that Triumph purists hated the TR7. Not only was it completely lacking in performance, but the handling was uninspiring as well. It used MacPherson struts at the front, and the rear end used a beam axle mounted on coil springs – the result was a chassis with little handling finesse, but very low build costs.

It was unanimously agreed that the TR7's styling didn't work. It was originally the work of Harris Mann, who also styled the Austin Allegro and Princess. But by the time the design had been passed through various committees, the TR7 had become shorter and taller, giving it an awkward appearance – which was not helped by the fact that it took until 1979 for a convertible to appear.

YUGO 45 *(1984–92)*

As a cheap car that was tidy, if not exciting to look at, the Yugo 45 and its derivatives could have had promise. Under the skin, it was all Fiat 127, using the same platform and engines. But it was appallingly built, and sporty models came with a hideous body kit that didn't even fit. Among the Yugo's more charming traits were window-winder handles that snapped off in your hands, a gearbox that wouldn't go into reverse unless you used both hands, and plastic fenders that looked like they were made out of metal – until they fell off.

However, the Yugo 45 was the cheapest car on offer in both Britain and the United States, and that meant it enjoyed a steady demand until the factory was demolished as Yugoslavia tore itself to pieces in the early 1990s.

SPECIFICATIONS

TOP SPEED:	127km/h (79mph)
0–96KM/H (0–60MPH):	21.6secs
ENGINE TYPE:	in-line four
DISPLACEMENT:	903cc (55ci)
WEIGHT:	765kg (1701lb)
MILEAGE:	6.7l/100km (42mpg)

Left: ***'Go new, go Yugo,' screamed the advertising department, but its cry fell on deaf ears. You could get a much better car for the money on the secondhand market.***

Power came from Fiat engines, but unfortunately they were coupled to a truly awful gearbox. Wear to the syncromesh was so common that, by the time the car had racked up about 82,000km (50,000 miles), the only way to get it into reverse was to switch the engine off, engage reverse gear, and start the engine again with the clutch depressed.

Inside, the quality of the materials was appalling. The plastics were awful and the window winders tended to snap off in your hands, switches fell off and the glove box opened at random, spilling its contents into the passenger footwell.

Under the skin, the Yugo 45 was almost identical to a Fiat 127. But parent company Zastava insisted on its own body style, and the finished result was all right. Until, that is, you moved into an upmarket model, and were forced to drive around with a silly body kit attached.

Handling was terrible – the car sat higher than a Fiat 127 to cope with poor road surfaces in its home country, which meant that it suffered from dreadful understeer when cornered at any speed. The bouncy ride and over-light steering did little to improve matters.

ZAZ ZAPOROZHETS (1967–90)

The Zaporozhets, or ZAZ 966, to give it its Soviet title, debuted in 1967 and looked like a clone of the German NSU Prinz – but there the similarities ended. Built in the Ukraine, the unfortunate 'Zappo' has more than once been referred to as the worst car ever built – and not without good reason. The handling was awful and performance was shocking, the gear lever felt and sounded like a pneumatic drill if you tried to change down, and none of the panels fitted properly. To add insult to injury, the Zappo's air-cooled engine clattered and was prone to seize at speed, thanks to oil starvation, while the simple cam-and-peg steering mechanism had a habit of collapsing as it aged. Driving a Zappo is truly, truly terrifying …

SPECIFICATIONS

TOP SPEED:	112km/h (70mph)
0–96KM/H (0–60MPH):	no figures available
ENGINE TYPE:	V4
DISPLACEMENT:	887cc (53ci)
WEIGHT:	no figures available
MILEAGE:	no figures available

Left: ***Apparently the Zappo was small, comfortable and economical. Well, the advertising people were right on one count – it wasn't very large …***

Styling wise, the Zaporozhets is actually quite pretty – but don't let its baby-Corvair appeal seduce you. The panels are a bad fit, it's prone to rust and it's horrible, very horrible, to drive.

The Zaporozhets uses an engine completely unique to the model. It's an air-cooled V4 unit that may sound terrible, but is actually amazingly tough. Note the large scoops found on the car's side, which force-fed air into the fuel system.

It's quite possible that the Zaporozhets is the worst-handling car of all time. Its platform used the cheapest possible build methods, which meant it got torsion-bar front suspension, coils and a beam axle at the rear, and slow-to-react worm and peg steering.

The Zappo did have one unique innovation. The passenger side floorpan was removable, so that owners could use it to go fishing on frozen lakes!

S NN 1830

DESIGN DISASTERS

Some cars are so dreadful they should never have made it past the initial design stage. But desperate times call for desperate measures, and that means some truly awful machinery has slipped through the net when car makers have realized their project is so far advanced that they can't afford to go back to the drawing board. The cars in this section are design disasters for many different reasons. Some are just fundamentally terrible designs, flawed in a number of ways and often lacking an element that would seem far too obvious for most people to omit. Why didn't the Austin Princess have a hatchback, for example? And why did the FSO Polonez come with a hatch, but no fold-down rear seat? These are questions that have never been answered.

Other entries are disastrous because they lacked mechanical detail. They were fairly adventurous, but failed because they hadn't been properly thought out. Then there are those whose design faults are instantly evident – if it's true that a car's design appeal starts with its styling, travesties such as the AMC Gremlin, Pontiac Aztek and Subaru XT Coupé should never have come into being …

Left: *The Mercedes Vaneo – proof that even today, car makers can get it horribly wrong sometimes.*

AMC GREMLIN (1970–79)

The early 1970s were disastrous for the American Motor Corporation, with an outdated model range and no money in the coffers to develop new cars. So when, in 1970, AMC launched a car to cope with fuel crises, it needed to be good to ensure the company's survival. It wasn't – and it marked the start of a steep downward spiral for one of America's greatest marques.

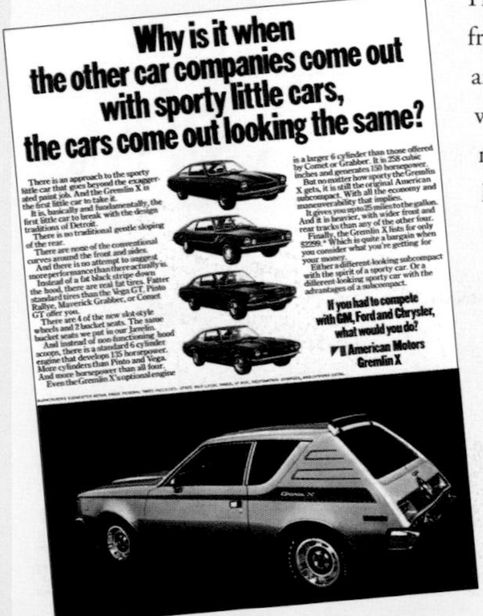

The Gremlin was hideously ugly from whatever angle you viewed it, and was also appallingly built, while the supposed 'hatchback' rear end was little more than a hinged window. AMC made its debut even worse by choosing to launch it on April Fools' Day. And although it was designed for economy, the Gremlin's 199ci six-cylinder engine wasn't exactly thrifty, returning about 25mpg. A V8-powered Gremlin X appeared the following year, and was even more embarrassingly bad.

SPECIFICATIONS

TOP SPEED:	153km/h (95mph)
0–96KM/H (0–60MPH):	12.5secs
ENGINE TYPE:	in-line six
DISPLACEMENT:	3802cc (232ci)
WEIGHT:	1278kg (2841lb)
MILEAGE:	14.1l/100km (20mpg)

Left: ***'Why is it that when other car companies come out with pretty little cars, the cars come out looking the same?' We don't know, but they're far better looking than the Gremlin!***

It might have looked like a hatchback, but it was only the Gremlin's rear window that opened – that was, if it didn't break off. Getting shopping from the bottom of your trunk could put your back out …

As if the Gremlin's debut weren't disastrous enough, dire build quality meant that it wasn't even reliable or rust-resistant. Sills and rear subframes were especially prone to rot.

AMC used its own six-cylinder engines in the Gremlin, and although they were of a fairly large capacity, performance was terrible thanks to emissions restrictions.

AMC PACER *(1975–81)*

The Pacer was the car that would eventually define AMC's breakdown as a company. It was launched in 1975 as an answer to the numerous Japanese and European models taking the American market by storm, and AMC had high hopes for its success. But it was grotesque to look at – too short, far too wide and blessed with a profile that made it look as if the front and back had been designed by two entirely different people. One door was longer than the other, supposedly to encourage owners to allow their children to access the back on the passenger side – but this feature was lost in right-hand-drive markets. Reliability was dreadful, it rusted instantly and, despite having a 4.2-litre engine and 16.6l/100km economy, it was dreadfully slow.

SPECIFICATIONS

TOP SPEED:	142km/h (88mph)
0–96KM/H (0–60MPH):	15.8secs
ENGINE TYPE:	in-line six
DISPLACEMENT:	4229cc (258ci)
WEIGHT:	1554kg (3455lb)
MILEAGE:	(18mpg)

Left: ***AMC went to great lengths to sing the Pacer's praises, but its claims fell on deaf ears. Despite a huge marketing budget and committed export campaign, the Pacer was an unmitigated flop.***

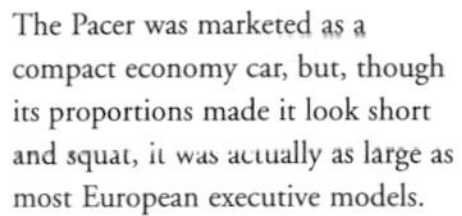

The Pacer was marketed as a compact economy car, but, though its proportions made it look short and squat, it was actually as large as most European executive models.

You could have a six-cylinder in-line engine or a V8, but it didn't matter which one you chose: both were slow and drank fuel like it was going out of fashion.

On left-hand-drive Pacers, the right-hand door was bigger to allow better access to the rear seats from the kerbside – a feature that was totally lost when AMC tried right-hand-drive exports.

AUSTIN 3-LITRE (1967–70)

Quite why the Austin 3-Litre ever existed is a complete mystery. At launch, everyone said it was an utterly pointless car, including some of the management at British Leyland. But the design team pressed ahead, going to great lengths to modify the platform of the front-wheel-drive Austin 1800 to accept a longer wheelbase and rear-wheel-drive transmission. Like most BL cars of its era, it suffered from rampant corrosion, while the 3.0-litre straight-six engine, modified from the aged Austin-Healey 3000, was prone to premature failure. Handling was awful, and the ride was poor for an executive car. Worse still, it was marketed as a competitor to cars such as the Jaguar XJ6, despite the fact that the 3-Litre had absolutely no image. After four years, Austin had managed to shift fewer than 10,000 units, so the decision was finally taken to put the 3-Litre out of its misery.

SPECIFICATIONS

TOP SPEED:	161km/h (100mph)
0–96KM/H (0–60MPH):	15.7secs
ENGINE TYPE:	in-line six
DISPLACEMENT:	2912cc (178ci)
WEIGHT:	1190kg (2645lb)
MILEAGE:	14.8l/100km (19mpg)

Left: *This woman is so ashamed she chose an Austin 3-Litre over a Jaguar that she's trying to hide inside her fur coat so nobody will recognize her.*

The majority of 3-Litres were broken for spares by Austin-Healey enthusiasts. The car used the same engine as the Austin-Healey 3000, but scrap cars were obviously much cheaper …

Some 3-Litre parts were scarce after the model had been out of production for just three years, including its unique propshaft and gear linkages.

Bearing in mind how much money BMC spent on the 3-Litre, you might have thought it started from scratch. But it used the doors and a heavily modified platform from the Austin 1800.

AUSTIN AMBASSADOR (1982–84)

It didn't take long for British Leyland to realize that the original Princess should have been given a hatchback rather than a conventional trunk, and a redesign started within months of the car going into production. It finally arrived in showrooms in 1982 as the Austin Ambassador, complete with an aerodynamic front end and folding rear seats to accommodate long loads. To cut costs, it used the same platform, doors and suspension system as the Princess, and shared its distinctive wedge-shaped profile.

But this was too little too late. By the time it arrived, the Ambassador was a dead duck. It was outdated compared to most modern rivals, and the atrocious 1.7-litre and 2.0-litre O-Series engines were notoriously fragile and started burning oil at less than 32,200km (20,000 miles). Top of the range was a Vanden Plas model, with a sumptuous interior and a unique radiator grille – a cynical attempts to sell it as an executive car.

SPECIFICATIONS

TOP SPEED:	161km/h (100mph)
0–96KM/H (0–60MPH):	14.3secs
ENGINE TYPE:	in-line four
DISPLACEMENT:	1994cc (122ci)
WEIGHT:	1203kg (2675lb)
MILEAGE:	10.9l/100km (26mpg)

Left: ***'Expensive motor cars in all but price.' And build quality. And style. And performance. And reliability. And image. Need we go on?***

Two power units were offered in the Ambassador, either a 1.7-litre (104ci) or 2.0-litre (122ci) O-Series overhead cam engine. Both were unrefined and dreadfully unreliable.

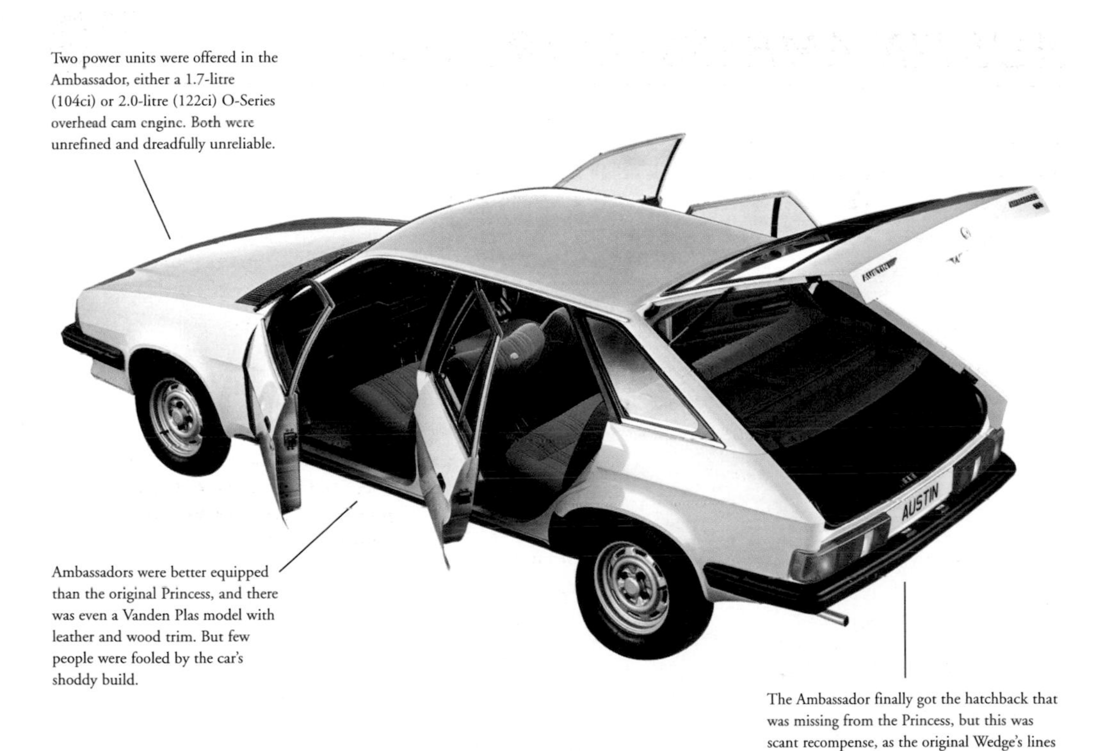

Ambassadors were better equipped than the original Princess, and there was even a Vanden Plas model with leather and wood trim. But few people were fooled by the car's shoddy build.

The Ambassador finally got the hatchback that was missing from the Princess, but this was scant recompense, as the original Wedge's lines were ruined by the new look.

AUSTIN PRINCESS 18-22 SERIES (1976–82)

In styling terms, the Princess is one of those shapes you either love or loathe. It soon became known as 'The Wedge', thanks to its distinctive trapezoid silhouette, while British Leyland went to great lengths to confuse buyers, offering it initially as an Austin, Morris or Wolseley, then dropping the lot and reverting to simply the 'Princess' name.

In fairness, the Princess was reasonably pleasant to drive, with acres of passenger space and a pleasant ride. But these were cold comfort for buyers, who soon got used to collapsing suspension units, shearing drive shafts and engine mounts that randomly fell out. Range-topping 2.2-litre versions had a lovely engine note, but the standard four-speed manual transmission was dreadful, and early cars were prone to excessive body rot. Then there was the most notable fault – the complete lack of a hatchback rear end. It was initially designed to include this feature, but BL dropped the plan so as not to steal sales from the struggling Maxi.

SPECIFICATIONS

TOP SPEED:	155km/h (96mph)
0–96KM/H (0–60MPH):	14.9secs
ENGINE TYPE:	in-line four
DISPLACEMENT:	1798cc (110ci)
WEIGHT:	1150kg (2557lb)
MILEAGE:	11.8l/100km (24mpg)

Beautifully thought out.
Beautifully made.

Left: ***The UK's Trade Descriptions Act was yet to come into force when British Leyland described the Princess as 'Beautifully thought out, beautifully made'. The company would have been fined, otherwise.***

The cabin of a Princess pretty much sums up the 1970s: it's full of naff gadgets, beige velour and hastily glued-on fake wood.

The range-topping six-cylinder model was smooth and reasonably powerful, but the gearbox was housed in the sump, so there was no way that British Leyland could offer a five-speed model to rival competitors.

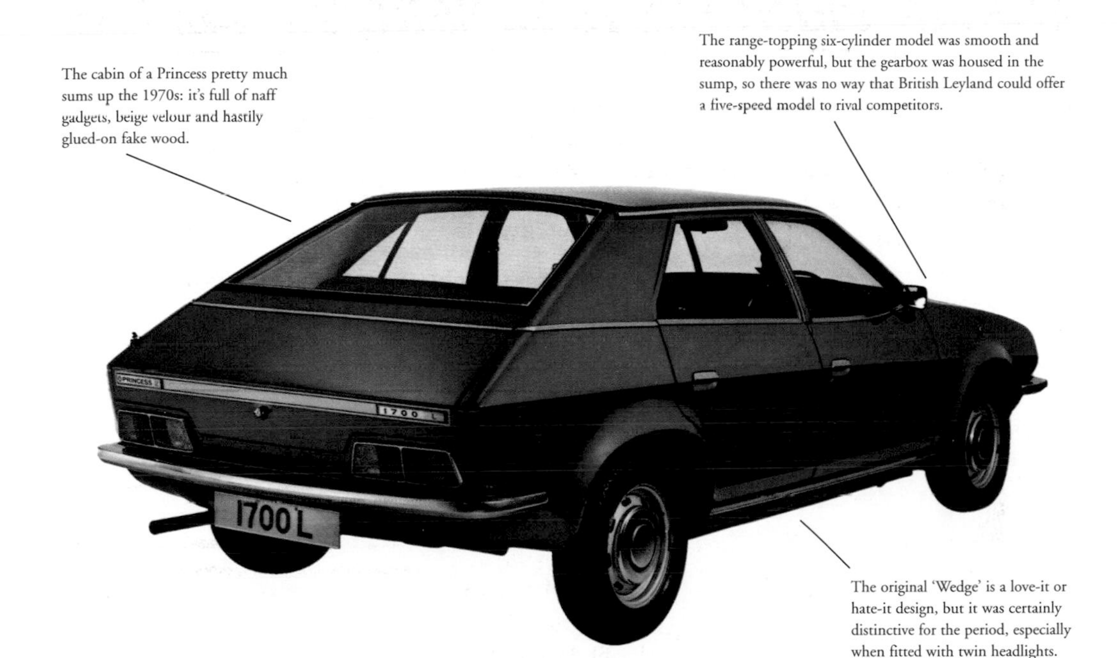

The original 'Wedge' is a love-it or hate-it design, but it was certainly distinctive for the period, especially when fitted with twin headlights.

CADILLAC SEVILLE *(1979–86)*

Styling is essential, especially in the luxury car market. But nobody appeared to tell Cadillac, as the US luxury maker laughed in the face of good taste when it unveiled this abomination in 1979. The Seville was overly adorned with chintzy false chrome trim and was also a ridiculous shape – motoring journalists at the time said it looked as if it had been reversed into a wall, or that perhaps the ceiling had collapsed in the styling studio and squashed the back end of the clay model. Whatever the reason, Cadillac soon regretted making the Seville look quite so daft, as buyers flocked away from Cadillac to far more tasteful European models, such as BMW and Mercedes. Its one saving grace was its impressive standard equipment.

SPECIFICATIONS

TOP SPEED:	187km/h (116mph)
0–96KM/H (0–60MPH):	10.4secs
ENGINE TYPE:	V8
DISPLACEMENT:	6054cc (369ci)
WEIGHT:	1982kg (4405lb)
MILEAGE:	20.2l/100km (14mpg)

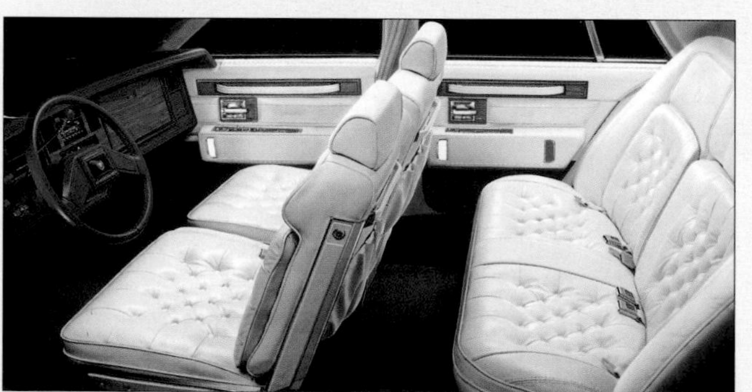

Left: ***There were toys aplenty inside the 1985 Seville, enough to keep even the most committed gadget-craver entertained, but as this illustration shows they were all in the worst possible taste!***

The Seville was engineered to give the ultimate in ride smoothness, but it was so soft that the handling became perilous if cornered at any speed, with no front-end grip.

The Seville was well equipped, but it wasn't what executive car buyers wanted. The cabin was trimmed in cheap-feeling plastics, and the best European rivals did it all so much better.

The tail end is definitely the talking point of the Seville because it falls so completely out of harmony with the rest of the car's design. Cadillac never did supply a reason for its stunted appearance.

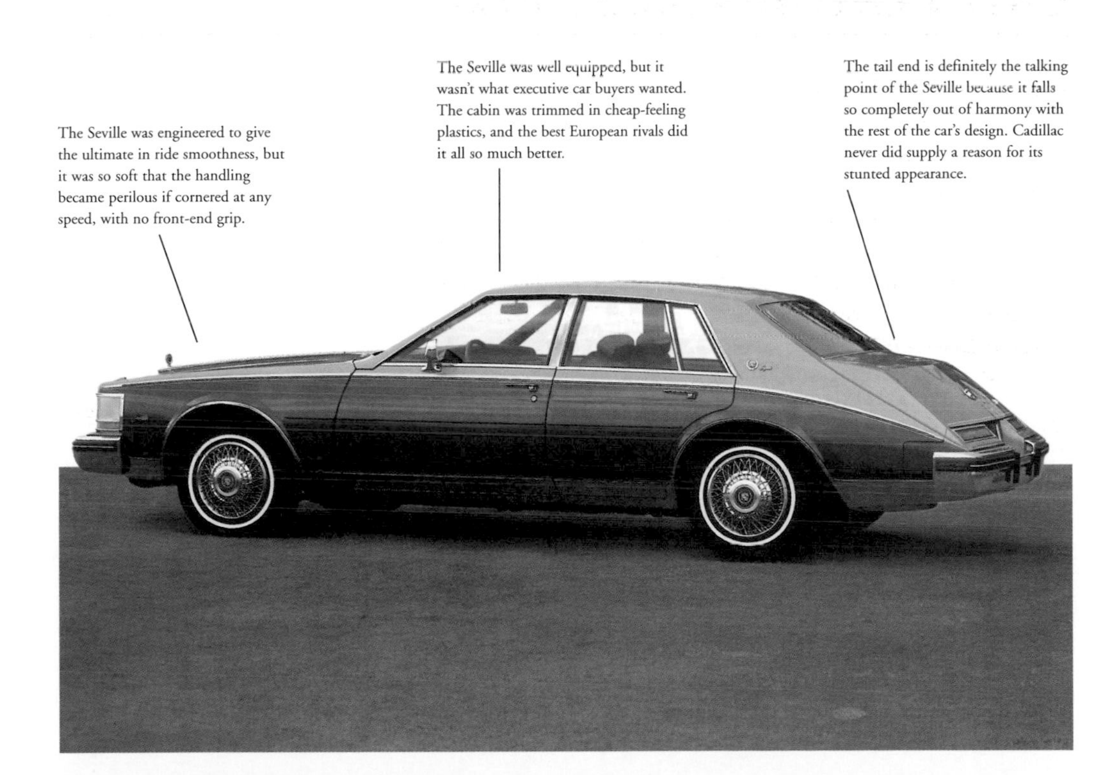

CADILLAC SEVILLE *(1997–2003)*

So confident was Cadillac that the 1998 Seville would be a success that it decided to launch the new model in Europe, as well as its established US market. But while the newcomer actually looked comparatively classy, it was the same old story beneath the skin. The cabin was comfortable, but had nothing in the way of style or build quality, with cheap, nasty plastics and a lack of any damping to the controls. It wasn't what European luxury car buyers wanted. Even in America, where Cadillac enjoyed a loyal customer base, many had switched to European brands with their better build. The Seville also lacked any kind of dynamic ability. The ride was soft and bouncy, and if you attacked a corner at any speed the car would just plough on in a straight line.

SPECIFICATIONS

TOP SPEED:	210km/h (130mph)
0–96KM/H (0–60MPH):	7.8secs
ENGINE TYPE:	V8
DISPLACEMENT:	4565cc (278ci)
WEIGHT:	1810kg (4022lb)
MILEAGE:	10.9l/100km (26mpg)

Left: ***According to Cadillac, the 1998 Seville was 'built to set the standard of the world' – conveniently forgetting that most German models were so much better than its car in almost every respect.***

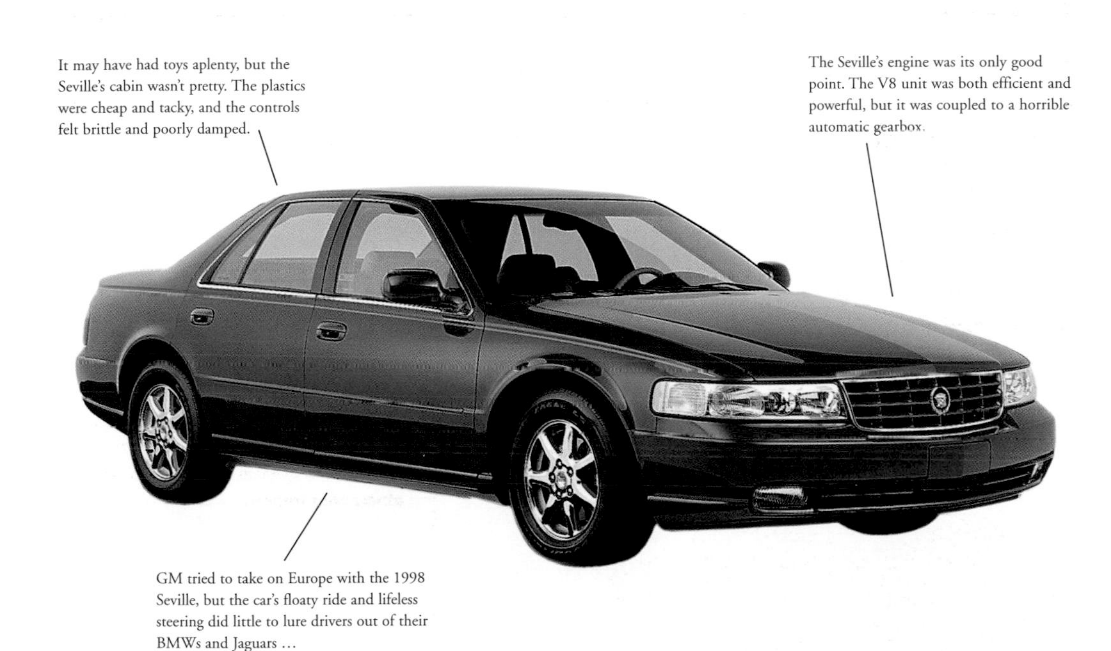

It may have had toys aplenty, but the Seville's cabin wasn't pretty. The plastics were cheap and tacky, and the controls felt brittle and poorly damped.

The Seville's engine was its only good point. The V8 unit was both efficient and powerful, but it was coupled to a horrible automatic gearbox.

GM tried to take on Europe with the 1998 Seville, but the car's floaty ride and lifeless steering did little to lure drivers out of their BMWs and Jaguars …

CHEVROLET CORVAIR *(1961–70)*

If one bad review is enough to damage a car's reputation, a whole book about its faults is enough to kill it. And that's exactly what happened to the Chevy Corvair. The rear-engined economy car was planned as General Motors' answer to budget models such as the VW Beetle. It was neatly styled and good value for money, and like the Beetle it was air-cooled. But sadly, it wasn't properly engineered; if the tyre pressures were at all wrong, it demonstrated a worrying amount of oversteer and a tendency to spin out of control at the merest provocation. Its downfall came in 1966, when Ralph Nader published a book called *Unsafe at Any Speed*, highlighting the Corvair's fundamental balance problems. Some of his findings were a little unfair, but the damage was done, and the Corvair never shook off its tainted image.

SPECIFICATIONS

TOP SPEED:	135km/h (84mph)
0–96KM/H (0–60MPH):	21.4secs
ENGINE TYPE:	flat-six
DISPLACEMENT:	2295cc (140ci)
WEIGHT:	1236kg (2720lb)
MILEAGE:	10.0l/100km (28mpg)

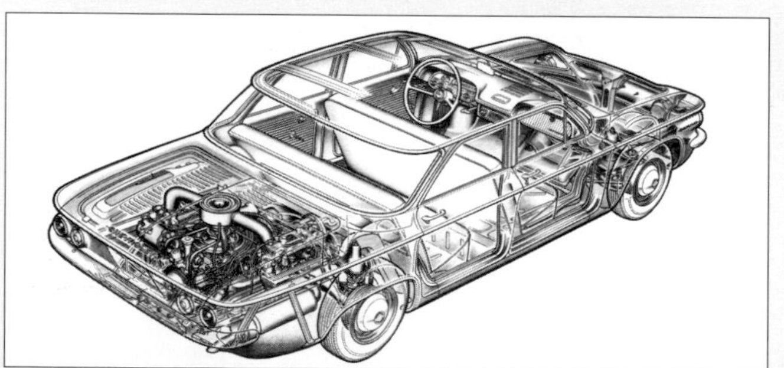

Left: ***The cutaway drawing clearly shows the source of the Corvair's handling issues. You can clearly see the swing axles, but also note the location of the engine, positioned way behind the rear axle.***

If there was one good thing about the Corvair, it was that it offered an exceptional degree of cabin comfort. The long wheelbase and lack of a transmission tunnel made for plenty of space …

The Corvair aimed to take on the VW Beetle with its air-cooled rear mounted engine, which was enjoying unprecedented success in the US market, but its motor was mounted too far back and as a result the chassis had no balance.

American lawyer and consumer rights advocate Ralph Nader's criticism of the Corvair focused on the rear suspension. The swing axle caused the rear wheels to tuck underneath the car if the tyre pressures were wrong.

CHRYSLER AIRFLOW (1934–37)

Sometimes, being ahead of the game isn't a good thing. The Chrysler Airflow is proof positive: much of its avant-garde engineering would later turn out to be commonplace in the world of car design. But in 1934, when the Airflow debuted, buyers weren't ready for such a streamlined shape. The car was designed using a wind tunnel and employing the aerodynamic expertise of air travel pioneer Orville Wright. Many aircraft influences were evident in its design, including a curved windshield and a lightweight body built on trusses. So, styling apart, why was the Airflow such a failure? The simple answer is that it was poorly executed, with clumsy build-quality problems, rust issues and a few too many mechanical problems. Buyers simply weren't prepared to pay the premium Chrysler was demanding over lesser models for an obscurely styled and unproven car.

SPECIFICATIONS

TOP SPEED:	142km/h (88mph)
0–96KM/H (0–60MPH):	19.5secs
ENGINE TYPE:	in-line eight
DISPLACEMENT:	4883cc (298ci)
WEIGHT:	1894kg (4166lb)
MILEAGE:	17.7l/100km (16mpg)

Left: *Viewed in a contemporary context, it's easy to see why the Airflow was a disaster: it just isn't in keeping with the exquisite details of 1930s style ...*

The view out was like that from an aeroplane, with a curved windshield and a front end that drooped steeply away from the cabin.

Even though the Airflow was aerodynamic, it wasn't economical. It weighed so much that at best the engine managed an average of only 17l/100km (16mpg).

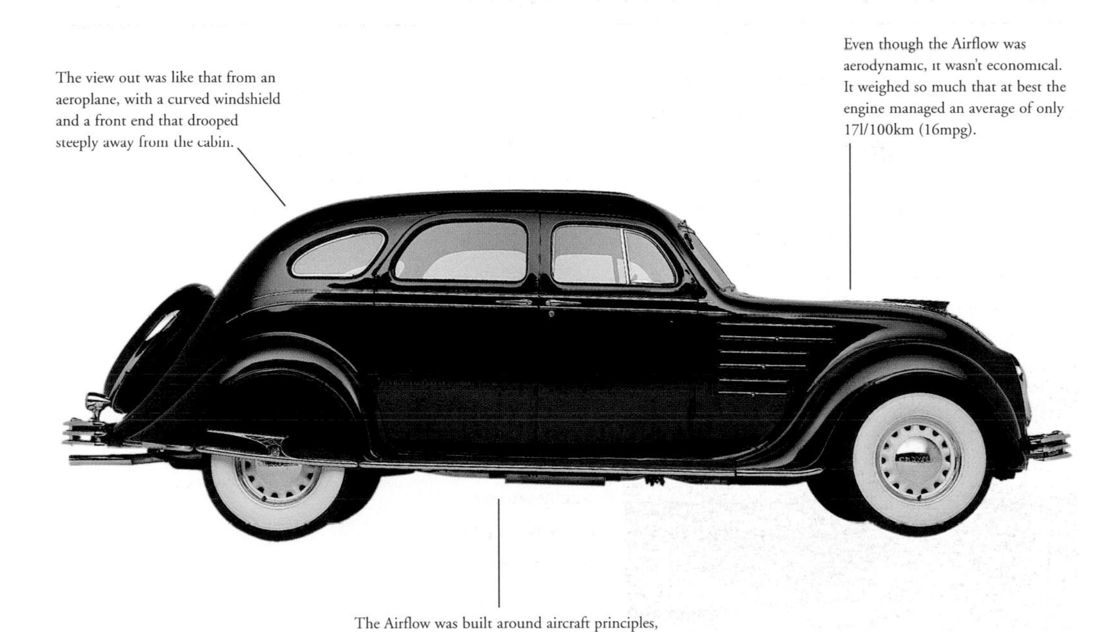

The Airflow was built around aircraft principles, with a view to creating the least resistant airflow around the car. The aerodynamics worked, but came at the expense of the styling!

CITROËN GS BIROTOR *(1974–75)*

Given its reputation for unusual and creative design, it's hardly surprising that Citroën was one of the first manufacturers to jump on the rotary engine bandwagon. It joined forces with German firm NSU to develop a range of twin rotary 'Wankel' engines in 1967, but the relationship didn't last long and the company – Comotor – was disbanded soon afterward. Citroën persisted, though, and the GS Birotor finally arrived in 1974. It was an enterprising idea – but ultimately a disaster. The GS Birotor was fast and stylish, but it was considerably more expensive than a four-cylinder variant and suffered from wear to the rotor seal tips, causing premature engine failure.

In the end, Citroën gave up on the project after selling just 847 examples. Most of them were bought back by Citroën and destroyed, as the French firm didn't want to deal with the issue of having to maintain a parts network.

SPECIFICATIONS

TOP SPEED:	175km/h (109mph)
0–96KM/H (0–60MPH):	14.0secs
ENGINE TYPE:	twin rotary Wankel
DISPLACEMENT:	2 x 497.5cc (2 x 30ci)
WEIGHT:	1170kg (2600lb)
MILEAGE:	12.8l/100km (22mpg)

Left: ***This diagram shows how the rotary engine worked – but wear to the rotor tips caused it to lose all compression.***

The Birotor was identical to the standard GS in terms of its styling – the only differences were a couple of extra cooling ducts in the front and discreet 'Birotor' badges.

Handling and performance were excellent thanks to the GS's light weight, but the rotary engine was inefficient and thirsty.

Citroën's tie-in with Comotor resulted in the GS being offered with rotary power, but as with all early rotary engines the unit was fragile and prone to premature failure.

EXCALIBUR SSK (1964–81)

The Excalibur SSK is the perfect car for the buyer with lots of money and not enough taste. Launched in 1964, it was one of the first ever 'nostalgia cars', which used the styling of a vintage model but incorporated modern running gear. The car was the brainchild of American industrial designer Brook Stevens, who adored the style of European racing cars. It was designed to be an exact replica of the Mercedes SSK – a car that collectors will happily pay millions for – but it turned out to be a poorly executed, overpriced kit car. It came with a Chevrolet Corvette engine and a fibreglass body, and later on it was offered as a four-door open tourer. But enthusiasts were horrified by the dishonesty of the concept behind Excalibur, and the car was also pretty nasty to drive. It was built for 15 years and just short of 2000 were built, bought mostly by eccentric millionaires.

SPECIFICATIONS

TOP SPEED:	201km/h (125mph)
0–96KM/H (0–60MPH):	7.0secs
ENGINE TYPE:	V8
DISPLACEMENT:	4738cc (289ci)
WEIGHT:	1957kg (4349lb)
MILEAGE:	18.8l/100km (15mpg)

Left: ***Even the Excalibur's badge tried to emulate the look of the original Mercedes-Benz, with the added touch that the famous three-pointed star was replaced by a sword to match the Excalibur name.***

It might have looked traditional on the outside, but the Excalibur was built for rich Americans. So a well-equipped interior and thick leather seats were mandatory.

Power came from a Chevrolet Corvette engine, and the side-exit exhausts were purely for show. Unlike those on the original Mercedes, which aided cooling …

The one-piece body moulding wasn't overly convincing. Made out of fibreglass, it didn't fit well on the car's bulky chassis.

FIAT MULTIPLA (1998–2004)

When it appeared in 1998, the Fiat Multipla caused quite a stir. Its styling challenged convention, and, although it was actually a decent people-carrier in most respects, its bizarre looks were enough to scare off the majority of buyers. Even Fiat was prepared to confess that the Multipla wasn't exactly beautiful – press demonstration cars in the UK came with a sticker in the rear window that read: 'Wait until you see the front!'

Motoring journalists raved about the Multipla's clever three-abreast seating arrangement, immense practicality and excellent driving position, but ultimately it was doomed to failure because buyers weren't brave enough to part with their cash. The Italian firm finally gave up with the Multipla's quirkiness in 2004, and restyled the nose and tail to give it a far more conventional, if less intriguing, appearance.

SPECIFICATIONS

TOP SPEED:	173km/h (107mph)
0–96KM/H (0–60MPH):	12.6secs
ENGINE TYPE:	in-line four
DISPLACEMENT:	1581cc (96ci)
WEIGHT:	1470kg (3266lb)
MILEAGE:	8.3l/100km (34mpg)

Left: Although it appeared much larger, the Multipla was based on the platform of the Bravo/Brava range of family hatchbacks, with the track increased to give greater width.

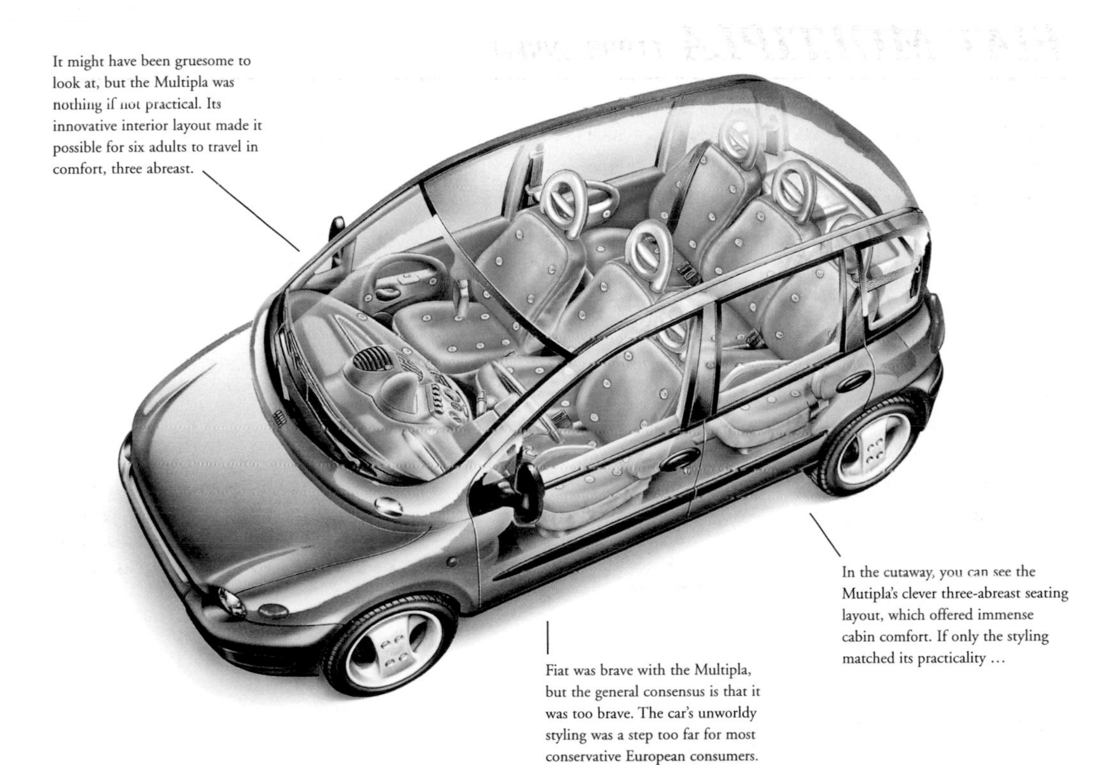

It might have been gruesome to look at, but the Multipla was nothing if not practical. Its innovative interior layout made it possible for six adults to travel in comfort, three abreast.

Fiat was brave with the Multipla, but the general consensus is that it was too brave. The car's unworldy styling was a step too far for most conservative European consumers.

In the cutaway, you can see the Mutipla's clever three-abreast seating layout, which offered immense cabin comfort. If only the styling matched its practicality …

FORD CLASSIC/CAPRI 109E (1961–64)

There was certainly nothing wrong with the styling of Ford's new family contender for 1961. The Classic bore several US design influences and looked fantastic, with its distinctive grille, quad headlights and reverse-raked rear window. The two-door Capri coupé looked even prettier, but the 109E series was a definite case of beauty being only skin deep. In Ford terms, the Classic and Capri were disasters – expensive to develop, difficult to build and slow sellers in a marketplace where buyers favoured more traditional designs. There were also problems with the original three-bearing engine, which was prone to premature big-end failure. And despite those alluring looks, neither was great to drive, with sluggish acceleration and heavy bodywork that caused the skinny tyres to lose grip prematurely. Production lasted just three years before the Cortina appeared, with far more success.

SPECIFICATIONS

TOP SPEED:	153km/h (95mph)
0–96KM/H (0–60MPH):	13.7secs
ENGINE TYPE:	in-line four
DISPLACEMENT:	1340cc (82ci)
WEIGHT:	898kg (1995lb)
MILEAGE:	10.0l/100km (28mpg)

Left: ***We haven't got a clue why there's a woman sitting on a deckchair in the trunk of this Capri 109E. Perhaps the driver ran out of seats inside?***

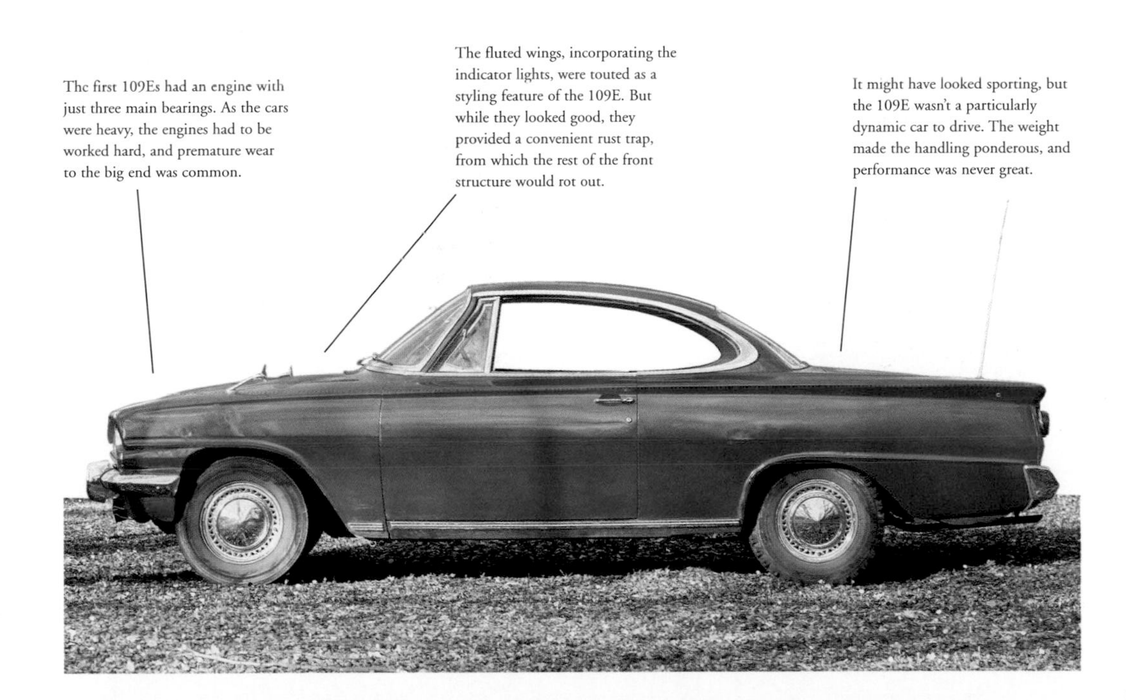

The first 109Es had an engine with just three main bearings. As the cars were heavy, the engines had to be worked hard, and premature wear to the big end was common.

The fluted wings, incorporating the indicator lights, were touted as a styling feature of the 109E. But while they looked good, they provided a convenient rust trap, from which the rest of the front structure would rot out.

It might have looked sporting, but the 109E wasn't a particularly dynamic car to drive. The weight made the handling ponderous, and performance was never great.

FORD ESCORT *(1990–93)*

Not since the Austin Allegro of 1973 was a car quite so lambasted by the European press as the 1990 Ford Escort. Its predecessor had enjoyed a nine-year production run and was considered one of the better cars in its class throughout, so buyers were expecting great things from the Blue Oval's newcomer. But it wasn't to be. Ford used the 1990 Motor Show in Birmingham, England, to unveil its replacement for the best-selling hatchback, and before journalists had chance to drive the car, critics were criticizing the styling, which was considered rather too dull to support the hype.

When the Escort hit the road and was found to be noisy, unrefined, poorly equipped, expensive and uninspiring, the press was savage. It took Ford just two years to launch a facelift, with spruced-up styling, new engines and massively improved dynamic capability.

SPECIFICATIONS

TOP SPEED:	179km/h (111mph)
0–96KM/H (0–60MPH):	11.9secs
ENGINE TYPE:	n-line four
DISPLACEMENT:	1597cc (92ci)
WEIGHT:	2224lb (1000kg)
MILEAGE:	32mpg (8.8l/100km)

Left: ***Ford's brochure for the 1990 Escort shows all of the body-style variants – three-door, five-door estate and cabriolet. There was also a four-door saloon, called the Orion.***

The Escort's credentials were not improved by Ford's decision to use an ageing engine range. The overhead valve entry-level models were tough but unrefined. Bigger CVH units were better, but weren't as economical as rivals.
Ford had a lot to live up to with the 1990 Escort, and it failed spectacularly. The car's most obvious failing was its insipid, bland styling.
The chassis was unimpressive, offering poor handling and an easily upset ride. The car was prone to understeer and the steering offered little feedback.

FORD MUSTANG MK II *(1973–80)*

What the Ford Motor Company thought it was playing at when it launched the Mk II Mustang is anyone's guess, but in doing so it completely devalued a brand name that had brought it multi-million dollar success over the previous decade. Whereas the original 1964 Mustang was billed as an instant classic – and quite rightly so – its 1973 replacement was an utter flop. Emaciated styling, enormous impact resistant fenders and a gaudy radiator grille did it no favours whatsoever, while it seemed that a special team had been drafted in to remove every last trace of driver appeal. Gone was the lively tail-happy chassis, replaced by a soggy, soft-sprung and incompetent suspension setup, while power outputs were reduced to comply with stringent US emissions legislation. A terribly sad demise.

SPECIFICATIONS

TOP SPEED:	156km/h (97mph)
0–96KM/H (0–60MPH):	13.0secs
ENGINE TYPE:	V6
DISPLACEMENT:	2798cc (170ci)
WEIGHT:	1382kg (3073lb)
MILEAGE:	14.1l/100km (20mpg)

Left: ***Who were they trying to kid? 'Mustang II, Boredom Zero.' Er, no: compared to the original classic Mustang, the Mk II was very, very dull indeed.***

Over-soft suspension and a crude rear-drive layout made the Mustang Mk II more of a boulevard cruiser than a driver's car, removing everything the original stood for.
The original Mustang was a design classic. Its successor was a travesty: it lacked any kind of harmony, the fenders were too big and the proportions were far too large.
Tough emissions legislation spelt the end of any performance. The engines were tuned down so far that the Mk II Mustang struggled to reach 160km/h (100mph), meaning that many European compact cars were quicker than the erstwhile American legend.

FORD PINTO (1971–80)

The Pinto was a PR disaster on a similar scale to the Chevrolet Corvair, with a massive recall issued on the grounds of public safety after it had been on the market for at least two years. The move came after several accidents involving Pintos being smacked in the rear: the pipe leading from the fuel filler to the tank would rupture and the car would burst into flames. There was a global outcry when it transpired that Ford had discovered the problem itself during its own testing, but had considered that the cost of modifying the fuel filler necks – about $50 a car – was too much for the company to instigate a redesign. After losing several lawsuits, Ford changed its tune; later cars were perfectly safe – but the damage had been done and the Pinto was always regarded with some suspicion.

Ford Pinto.
The best-selling small car in America.

FORD PINTO

FORD

SPECIFICATIONS

TOP SPEED:	131km/h (82mph)
0–96KM/H (0–60MPH):	18secs
ENGINE TYPE:	in-line four
DISPLACEMENT:	1599cc (98ci)
WEIGHT:	979kg (2176lb)
MILEAGE:	10.0l/100km (28mpg)

Left: ***Ford must have been worried when it discovered the Pinto was America's bestseller – it could have meant many lawsuits for the company!***

A fault with the fuel pipe meant that if a Pinto was bumped up the rear end, it could easily catch fire. Ford knew about the fault, but did nothing to rectify it until the lawsuits piled up.

The Pinto was an economy car, and it showed. The interior was trimmed in austere black plastic, and the car's equipment levels were basic, to say the least.

The Pinto's build quality did nothing to enhance its already poor reputation. Rust ate into the sills and valance panels.

FORD SCORPIO (1994–2000)

There was a notable shift in the European executive car market in the mid- to late 1990s, with buyers drifting away from mainstream brands and buying into those with a premium image, favouring models from Germany and Sweden. Ford's final bite at the Euro-executive cherry was the Scorpio, and it turned out to be an unmitigated disaster. Not that there was anything wrong with the way it drove: the big rear-drive saloon exhibited the dynamic excellence for which Ford became universally respected after the launch of the Mondeo in 1993. The build quality wasn't bad, either, equipment levels were generous and pricing was more than class competitive. So why did it fail? For the most obvious reason of all: it looked like a wide-mouthed frog and proved beyond reasonable doubt that, for car buyers, first impressions count …

SPECIFICATIONS

TOP SPEED:	222km/h (138mph)
0–96KM/H (0–60MPH):	9.0secs
ENGINE TYPE:	V6
DISPLACEMENT:	2933cc (179ci)
WEIGHT:	1581kg (3515lb)
MILEAGE:	10.0l/100km (28mpg)

Left: ***According to the advertisement, the Scorpio lives in another world. It's obviously a parallel universe, where people like cars that look like frogs.***

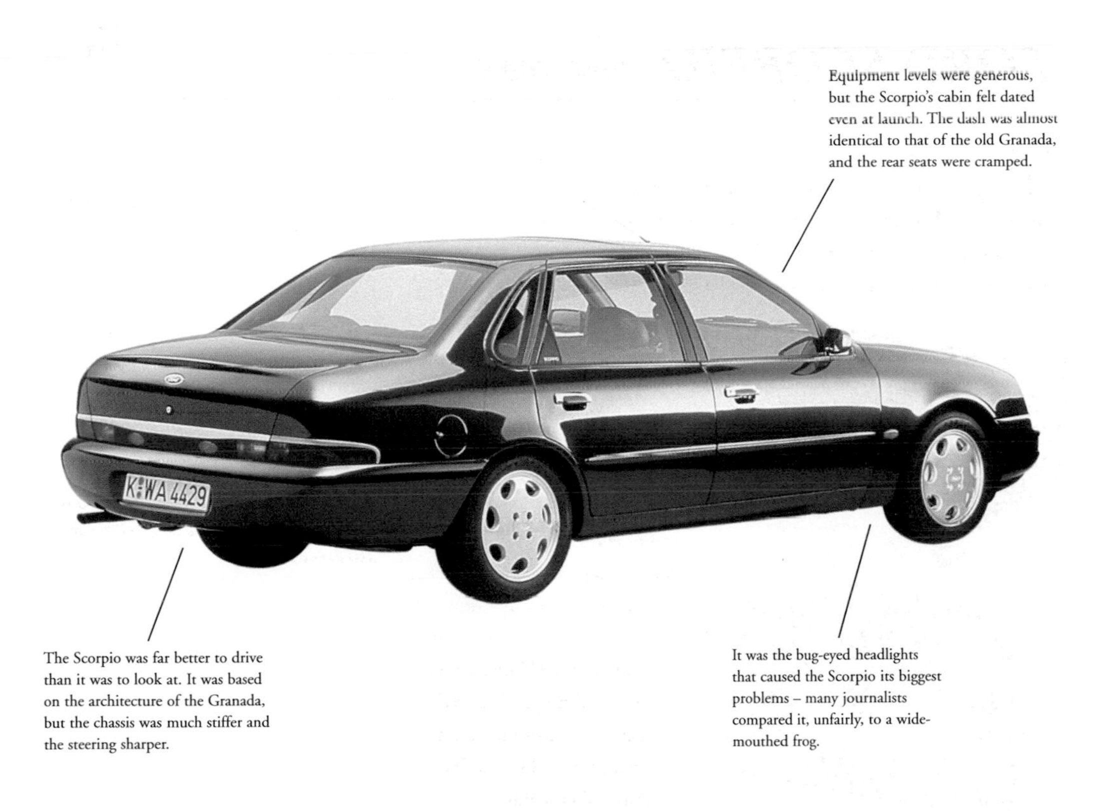

Equipment levels were generous, but the Scorpio's cabin felt dated even at launch. The dash was almost identical to that of the old Granada, and the rear seats were cramped.

The Scorpio was far better to drive than it was to look at. It was based on the architecture of the Granada, but the chassis was much stiffer and the steering sharper.

It was the bug-eyed headlights that caused the Scorpio its biggest problems – many journalists compared it, unfairly, to a wide-mouthed frog.

FSO POLONEZ/CARO (1985–98)

After Fiat withdrew from FSO on the grounds that its association with the Polish brand did its reputation few favours, the company was left to develop a new car – the Polonez. Despite entirely new bodywork, it still used an archaic suspension setup and a 1500cc Fiat engine, built under licence from the Italian maker. In styling terms, it wasn't pretty, even if its trapezoid profile and twin-headlight grille were in keeping with contemporary styling from other European makers. But it had precious little to recommend it. The handling was terrifying, the build quality was atrocious and the use of ultra-cheap paint meant that most Polonezes rotted away early on. Then there was that fundamental design flaw in a hatchback – the rear seat didn't drop. The car was briefly revived as the Caro in the early 1990s, this time using Peugeot engines, but it was hardly any better.

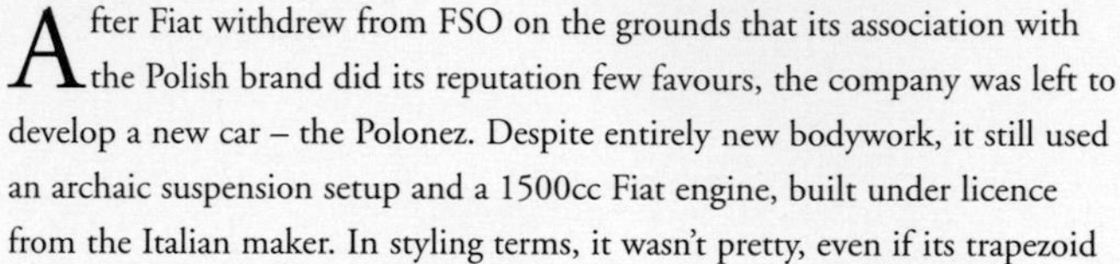

SPECIFICATIONS

TOP SPEED:	150km/h (93mph)
0–96KM/H (0–60MPH):	18.8secs
ENGINE TYPE:	in-line four
DISPLACEMENT:	1481cc (90ci)
WEIGHT:	1106kg (2459lb)
MILEAGE:	10.0l/100km (28mpg)

Left: ***It's sunset, and the owner of this Polonez still hasn't managed to get the damned thing started. Looks like he's here for the evening ...***

Cabins were awful – the design lacked any kind of logic, plastics were cheap, and although the Polonez had a hatchback, the rear seat couldn't be folded, leaving it with a tiny trunk.

FSO couldn't afford to develop its own engine, so the Polonez had an engine carried over from the Fiat 125 and dating back to the 1950s.

It was designed for rough Eastern European roads, so the Polonez sat high off the ground. That gave it dreadful balance problems on smoother Western European roads.

HILLMAN IMP (1963–76)

Had Hillman not left its customers to do the development work after the car went on sale, the Imp could have been one of the great success stories of the British motor industry. Sadly it wasn't, and the blame could be apportioned equally between the manufacturer and the British government. Rootes Group was given a substantial sum to invest in a new factory in Linwood, Scotland, where unemployment was rife, and a staff of ex-shipbuilders was brought in to assemble the Imp. Then workers got wind that their colleagues in Coventry earned more, and militant unionism meant that both build quality and development suffered. But Rootes had arranged a date for the first car to come off the line, with the Duke of Edinburgh booked to drive it, so the car was hurried into production unfinished and the earliest examples were shockingly unreliable.

SPECIFICATIONS

TOP SPEED:	130km/h (80mph)
0–96KM/H (0–60MPH):	25.4secs
ENGINE TYPE:	in-line four
DISPLACEMENT:	875cc (53ci)
WEIGHT:	688kg (1530lb)
MILEAGE:	7.6l/100km (37mpg)

Left: ***If test driving an Imp was 'an amazing experience that you owe yourself', then what must it have been like to drive an infinitely better car?***

Among its many clever features, the Imp had a hinged rear window that provided extra stowage space behind the rear seat. If only it had been better built, it could have been a success.

The spare wheel lived under the hood. With the engine at the rear, it gave weight to the front of the car and helped balance the handling.

It was the shoddy build quality that killed the Imp. As well as leaking water, the car rusted drastically, especially around the front firewall and door bottoms.

JAGUAR XJ40 *(1986–94)*

'A British Car to Beat the World.' That's how the XJ40 was billed when it made its Motor Show debut in October 1986, and on paper it should have been. A replacement for the XJ6 range had been on the drawing boards since 1972, but several setbacks stood in the way of the new model, and it took 14 years to reach showrooms, by which time it had morphed into what was allegedly the most technically sophisticated car on the planet. Certainly, it sounded impressive, with over three kilometres (1.8 miles) of wiring in every car, electronically aided self-levelling suspension and an on-board fault diagnosis system. It drove beautifully, too, with a fabulous ride. But sadly, it was underdeveloped. Early cars had terminal rust problems, and the on-board computer developed an odd habit of diagnosing faults that never existed. Later cars were genuinely good, but the earliest examples were horrors.

SPECIFICATIONS

TOP SPEED:	(235km/h) 141mph
0–96KM/H (0–60MPH):	7.4secs
ENGINE TYPE:	in-line six
DISPLACEMENT:	3590cc (237ci)
WEIGHT:	1653kg (3674lb)
MILEAGE:	16.6l/100km (18mpg)

Left: ***From the sublime to the ridiculous: the XJ40 started life as a true great, but ended up as something of a joke.***

Along with its luxurious interior, the XJ40 had an LCD display on the dashboard that warned of potential faults, but it nearly always gave phantom readouts. When there was a genuine problem, most owners ignored it …

All XJ40s came with anti-lock brakes as standard, but these proved frustrating for most owners as they packed up with alarming regularity. The sensors malfunctioned, causing the dash display to claim there was a fault and rendering the car unroadworthy.

Poor quality on early cars caused them to rust early, and the most vulnerable areas were the rear arches, front valance, sills and door bottoms.

LAMBORGHINI ESPADA *(1969–78)*

The Espada is nothing if not fascinating. When it debuted in 1969, it was the fastest four-seater production car in the world, with a ferocious V12 engine packed tightly in its nose. But power is nothing without control, and sadly the car had ridiculously heavy steering, meaning you needed the biceps of a shot-putter to be able to corner it any speed. On top of that, reliability was disastrous, with engines starving themselves of oil if driven too quickly, and build quality was fragile to say the least. The weirdness of its double-glass doors and droopy front end was beaten by the interior, which had the most haphazard switchgear arrangement imaginable – and after a few years hardly any of the switches worked, as corrosion in the earth points led to many an electrical malady. An intriguing curiosity, maybe. But definitely not a good car.

SPECIFICATIONS

TOP SPEED:	248km/h (155mph)
0–96KM/H (0–60MPH):	6.9secs
ENGINE TYPE:	V12
DISPLACEMENT:	3929cc (240ci)
WEIGHT:	1683kg (3740lb)
MILEAGE:	16.6l/100km (17mpg)

Left: *Lamborghini obviously didn't want any applause for the Espada, which is a good job, really, because it didn't get any.*

You had to be careful when opening an Espada door in a car park – the glass in the lower part of the panel would shatter if knocked.

There was a lot wrong with the Espada, but not under the hood! The car's amazing V12 powerplant was by far its most glorious feature.

There was no method or logic to the Espada's interior layout. The switchgear was all over the place, meaning you had to stretch across the cabin to locate the headlight switch!

LEXUS SC430 *(2001–present)*

SPECIFICATIONS	
TOP SPEED:	250km/h (155mph)
0–96KM/H (0–60MPH):	6.4secs
ENGINE TYPE:	V8
DISPLACEMENT:	4293cc (262ci)
WEIGHT:	1720kg (3822lb)
MILEAGE:	12.3l/100km (23mpg)

There's no denying that Lexus is one of recent motoring history's biggest success stories. The luxury arm of Toyota has produced some fine cars since its first appearance in 1989, but the SC430 most certainly wasn't one of them. It was supposed to be Lexus's answer to the Mercedes-Benz SL Series, but in reality it was a cynical attempt to combine the appeal of a luxury car with that of a sports car, without putting any attention to detail into the project. It even had an SL-style folding metal roof, but again this completely lacked the sense of occasion demonstrated by the German marque.

Inside, the SC430 was sumptuously equipped, but somehow it managed to completely lack any real sense of taste – an all-important factor in the luxury sector. Then there were its dynamic abilities – or, rather, its lack of them. Compared to established rivals, the Lexus's road behaviour was clumsy and, further, the steering lifeless.

Left: ***The only impressive thing about the SC430 was the way its roof folded away quickly and silently.***

It doesn't have many claims to fame, but one of the SC430's best features is its folding roof, which disappears into the deck at the touch of a button in less than 20 seconds.

Power comes from the Lexus LS430's V8 engine, which is far more suited to cruising than it is to sports car use. It's quick, but lacks top-end surge.

The key to making a good sports car is dynamic styling, but the SC430 looks far too round and fat to rival the likes of the Mercedes-Benz SL.

MERCEDES VANEO (2001–present)

In recent years, Mercedes has gone to great lengths to ensure that it has a car available in every imaginable market sector. And while some of these niches have been a remarkable success for the company – notably the supermini-sized A-Class and SLK Roadster – it has always taken shortcuts with its MPVs. The enormous van-based V-Class was bad enough, but the smaller Vaneo was a particularly poor attempt by Mercedes at cashing in on a lucrative market sector.

Even though it isn't a van, it looks like one, while the name does it no favours, implying that it is a commercial vehicle pretending to be a family car. The engines are underpowered and handling abilities are distinctively average, while inside the Vaneo seems to lack both the quality and integrity that have long been Mercedes hallmarks. It's not surprising, then, that the Vaneo's sales success proved somewhat limited.

SPECIFICATIONS

TOP SPEED:	180km/h (111mph)
0–96KM/H (0–60MPH):	11.1secs
ENGINE TYPE:	in-line four
DISPLACEMENT:	1598cc (92ci)
WEIGHT:	1290kg (2866lb)
MILEAGE:	8.3l/100km (34mpg)

Left: ***The Vaneo may look like a van from the outside, but the interior is as luxurious as any Mercedes-Benz. That's certainly no reason for buying one, though.***

Engines come from the Mercedes A-Class, and, while these are fine in a small car, they lack performance in the Vaneo and aren't especially efficient.

Despite it's name, the Mercedes Vaneo isn't actually based on a van. So why did the stylists make it look like one?

The Vaneo offers plenty of space, and there's even a seven-seater version. But it's far too expensive, and rivals from Citroën, Peugeot and Fiat do the job so much better.

MINI CLUBMAN *(1969–81)*

You can't improve on perfection, and the Mini Clubman is all the evidence you'll ever need. It was launched in 1969 as a supposed update to the then 10-year-old Mini, and, in creating the Clubman, British Leyland's designers took the original Alec Issigonis shape and modified it to wear the company's new corporate nose, shared with the unspectacular Maxi. In essence, it wasn't such a bad idea – after all, the original Mini's engine bay was cramped, which made it difficult to work on. But whoever was responsible for the redesign was unsympathetic towards the original Mini's gorgeous looks, and the Clubman appeared too long, while the squared-off nose sat awkwardly with the curvaceous tail inherited from the standard Mini. After 12 years of production, where it sold alongside the original car, the Clubman was shelved and the original continued.

SPECIFICATIONS

TOP SPEED:	145km/h (90mph)
0–96KM/H (0–60MPH):	13.3secs
ENGINE TYPE:	in-line four
DISPLACEMENT:	1275cc (78ci)
WEIGHT:	699kg (1555lb)
MILEAGE:	7.0l/100km (40mpg)

Left: ***While Britain may indeed be a small country, driving from one end to the other in a Mini Clubman remains a deeply uncomfortable experience.***

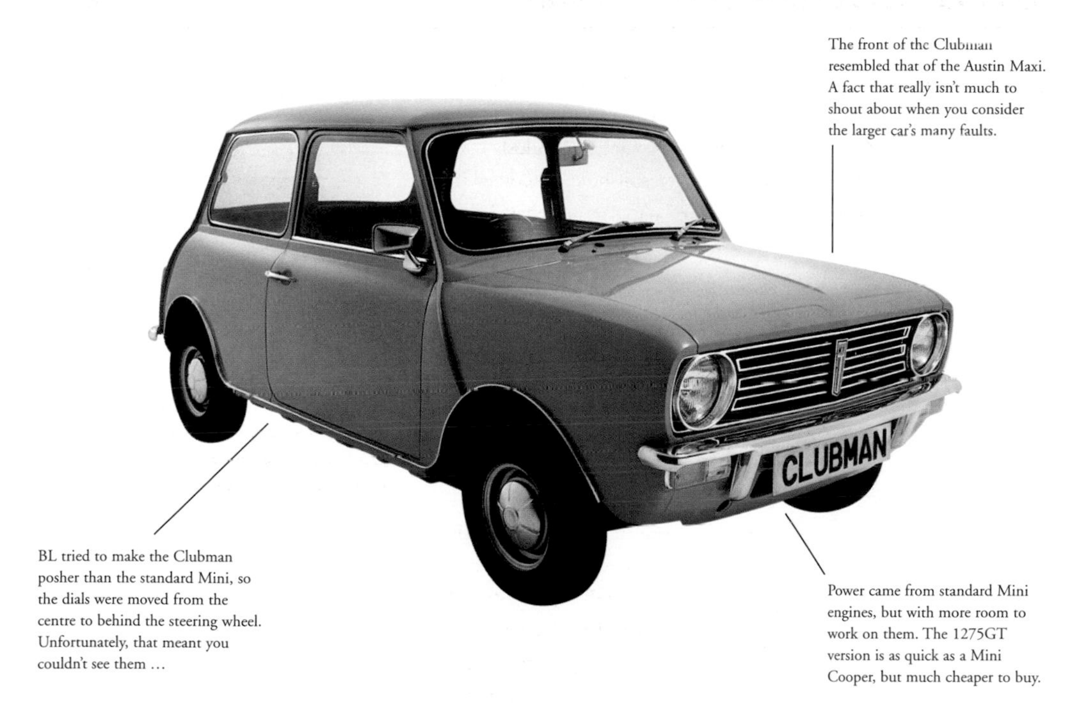

The front of the Clubman resembled that of the Austin Maxi. A fact that really isn't much to shout about when you consider the larger car's many faults.

BL tried to make the Clubman posher than the standard Mini, so the dials were moved from the centre to behind the steering wheel. Unfortunately, that meant you couldn't see them …

Power came from standard Mini engines, but with more room to work on them. The 1275GT version is as quick as a Mini Cooper, but much cheaper to buy.

MITSUBISHI CARISMA (1996–present)

Possibly the most inappropriately named car of all time, the Mitsubishi Carisma was about as uncharismatic as they come. For here was a car with no redeeming features whatsoever. It was anonymously styled, with a plain grey plastic cabin and horrible cloth seats. Under the hood, it used a gutless, noisy four-cylinder engine, coupled to a nasty, rubbery gearbox, and anyone in search of an even more depressing driving experience could choose a diesel engine that sounded like a bottle bank in an earthquake.

As if this were not enough to deter buyers, it also suffered from a complete lack of dynamic talent, suffering from plough-on understeer and a jittery ride. About the only good things you could say about the Carisma were that it had a big trunk and was generally reliable – but that does not make it a good car. Unless, of course, you're a cab driver …

SPECIFICATIONS

TOP SPEED:	200km/h (124mph)
0–96KM/H (0–60MPH):	10.4secs
ENGINE TYPE:	in-line four
DISPLACEMENT:	1834cc (112ci)
WEIGHT:	1754kg (3897lb)
MILEAGE:	7.2l/100km (39mpg)

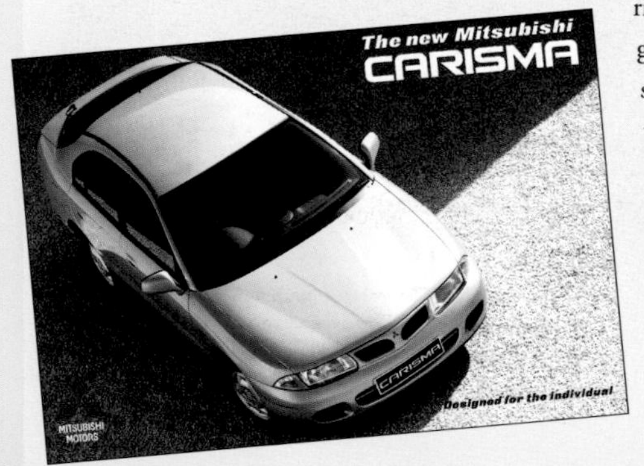

Left: ***Mitsubishi's advertising people were asking for it when they described the Carisma as 'designed for the individual'. Believe us – it wasn't.***

Mitusbishi makes reliable engines, but in the Carisma these are noisy and unrefined, while the gear change is rubbery and unpleasant to use.

Grey plastic, acres of it: the Carisma's interior is depressing in the extreme. And the hard cloth seats mean it isn't even comfortable.

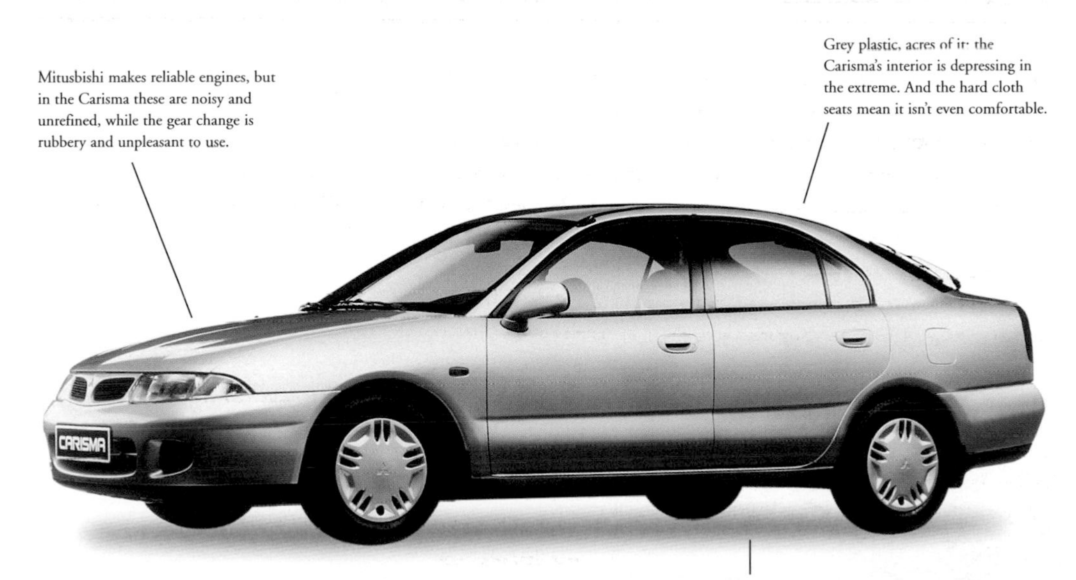

The styling is about as uninspiring as car design gets. Bland curves and a soulless profile make the Carisma utterly disenchanting, regardless of its name.

NISSAN SERENA *(1992–2001)*

If a prize were ever awarded for the dullest car ever built, the Nissan Serena would be in with a very good chance of winning the top honours. The Japanese company's mid-1990s MPV was supposed to take on the likes of the Renault Espace and Ford Galaxy, but in reality it was a truly shocking machine, handicapped with terrible handling, appalling styling and an interior that was greyer than a Siberian winter.

It was not even pretty – the Serena was based on the shell of the Nissan Urvan panel van, with some windows dropped in the side and six chairs nailed to the floor. But that wasn't the worst of the Nissan's problems. In normally-aspirated diesel guise, it was also the slowest car on sale in Europe, dawdling from 0–100km/h (0–62mph) in 34.8 seconds, if you could summon up the energy to grind your way through the industrial gearbox.

Nissan finally put it out of its misery in 2001, but its legacy lives on in the form of the LDV Cub – a panel van based on Nissan's aged Urvan design, and powered by the same clunky old diesel engine.

Left: One of the few redeeming features in the Nissan Serena was its great interior space – but you could also find this on much more impressive MPVs, such as the Ford Galaxy and Renault Espace.

SPECIFICATIONS

TOP SPEED:	132km/h (81mph)
0–96KM/H (0–60MPH):	34.8secs
ENGINE TYPE:	four-cylinder diesel
DISPLACEMENT:	1952cc (121ci)
WEIGHT:	1320kg (2955lb)
MILEAGE:	7.1l/100km (40mpg)

The most popular Serena was the 2.0-litre diesel, which brought a grin to the faces of sadistic taxi drivers, who knew it was so annoyingly slow, it would give following motorists road rage. If you craved more power, there was a 2.2-litre model that took 'only' 26.5 seconds to reach 100km/h (62mph).

At least the Serena was practical, with room for six and plenty of luggage space. But its van origins meant it had a high floor, so sitting in the back two rows of seats meant you ended up with legs akimbo, with your knees up near your shoulders.

It is a good job the Serena was too slow to challenge its chassis, for underneath it featured the least communicative steering since the Morris Marina, a beam rear axle and wheels that were far too small for its van-like body.

PONTIAC AZTEK *(1999–2004)*

Quite what Pontiac was thinking when it wheeled out this abomination is anyone's guess, but one thing's for certain – the Aztek is quite possibly the worst car produced by a mainstream American car manufacturer in recent years. Just look at it: the styling displays a complete lack of harmony, which would be amusing in a toddler's colouring book, but in a car intended to take on the burgeoning MPV market sector, it's a visual disaster. It looks like it has been made out of Lego.

And things don't get much better once you've climbed on board, as the interior is trimmed in horrid, low-quality plastics and the driving position is awful. Not surprisingly, the Aztek hasn't been the sales success that Pontiac hoped for, and is considered something of a joke among US motoring writers.

SPECIFICATIONS

TOP SPEED:	180km/h (111mph)
0–96KM/H (0–60MPH):	12.4secs
ENGINE TYPE:	V6
DISPLACEMENT:	3350cc (204ci)
WEIGHT:	1835kg (4077lb)
MILEAGE:	12.8l/100km (22mpg)

Left: ***Whichever angle you looked at it from, the Aztec was an ugly car. However, from the rear its chief saving grace is apparent – the practicality of plenty of accessible storage space.***

One glance at the Aztek, and you can see plainly why it's considered a bad car. The body is an incongruous mishmash of curves, flat surfaces and strange angles, and none of it works.

It might be distinctive from the outside, but the cabin of the Aztek is as plain and boring as they come. The dashboard could be from any Pontiac, while the driving position is distinctly uncomfortable.

The Aztek's faults could be overlooked if it were good to drive, but it's not. The handling is fairly wayward, the ride far too bouncy and the interior too cramped for comfort.

RENAULT AVANTIME *(2002–2003)*

There's something quite alluring about the Renault Avantime, despite it being remembered as one of the biggest flops of recent motoring history. The car was doomed to failure by its very concept. Using the platform of the Renault Espace people carrier, the French manufacturer decided to build a coupé with a hint of the Espace in its styling. What finally arrived after four years planning was utterly pointless.

It was huge on the outside, yet barely big enough inside to seat four people. The two-door bodyshell used enormous doors, which were too big to open in a tight parking space, and the out-of-this-world styling turned many a head, but not necessarily for the right reasons. When coachbuilder Matra, which built the Avantime for Renault, went bust in 2003, the model was canned for good – after less than a year in production. A disaster, though in design terms, a very brave effort.

SPECIFICATIONS

TOP SPEED:	222km/h (138mph)
0–96KM/H (0–60MPH):	8.6secs
ENGINE TYPE:	V6
DISPLACEMENT:	2946cc (180ci)
WEIGHT:	1741kg (3868lb)
MILEAGE:	11.3l/100km (25mpg)

Left: ***The Avantime certainly isn't large once you climb inside, even if it does look large. There's only room for four, and those in the back need short legs …***

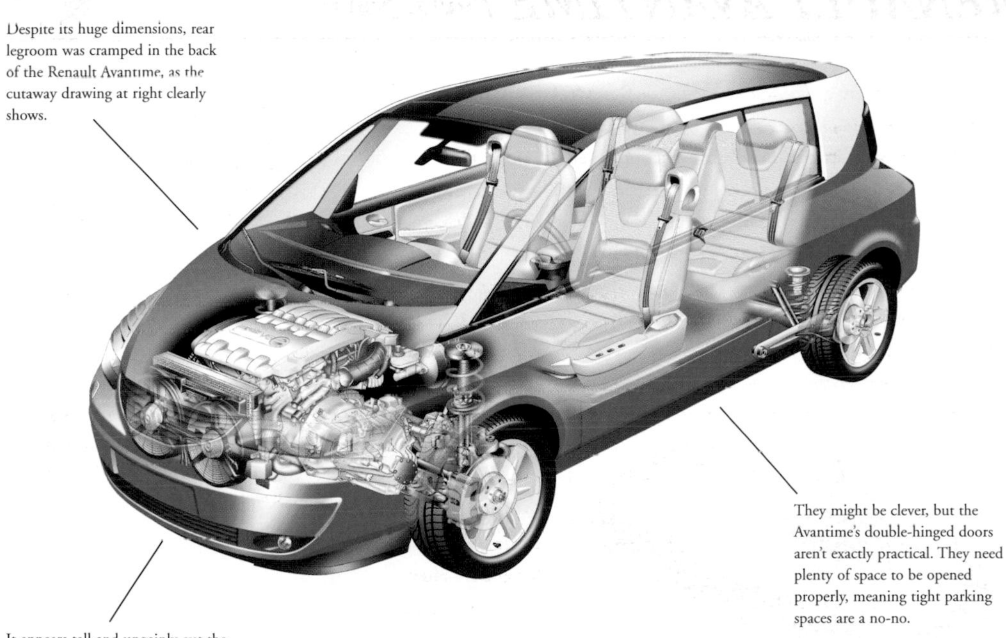

Despite its huge dimensions, rear legroom was cramped in the back of the Renault Avantime, as the cutaway drawing at right clearly shows.

They might be clever, but the Avantime's double-hinged doors aren't exactly practical. They need plenty of space to be opened properly, meaning tight parking spaces are a no-no.

It appears tall and ungainly, yet the Avantime's lightweight fibreglass bodywork makes for surprisingly agile handling and a low centre of gravity.

VAUXHALL VECTRA (1995–2002)

General Motors of Europe promised great things when it came to replace the ageing Vauxhall Cavalier range in 1995. Expectant customers were waiting for a new family model that would stun the world, but what they were given was a damp squib of a car that brought nothing new to the market other than an aerodynamic pair of door mirrors. It even looked like the car it was supposed to replace – when a Vectra was parked next to a Cavalier, it was difficult to tell the two apart in profile.

Not only that, but the Vectra's arch rival, the Ford Mondeo, was a truly brilliant car in its class and, despite having been on sale for two years already, was a vastly superior model. The Vectra gradually improved as time went by, but the earliest cars suffered from coarse engines, dull handling and an unnecessarily hard ride, along with a complete lack of flair.

SPECIFICATIONS

TOP SPEED:	208km/h (129mph)
0–96KM/H (0–60MPH):	8.7secs
ENGINE TYPE:	in-line four
DISPLACEMENT:	1998cc (122ci)
WEIGHT:	1369kg (3042lb)
MILEAGE:	10.0l/100km (28mpg)

Left: ***The Vectra's cabin may have been comfortable, but it was an ergonomic disaster. The electric window switches were down by the handbrake.***

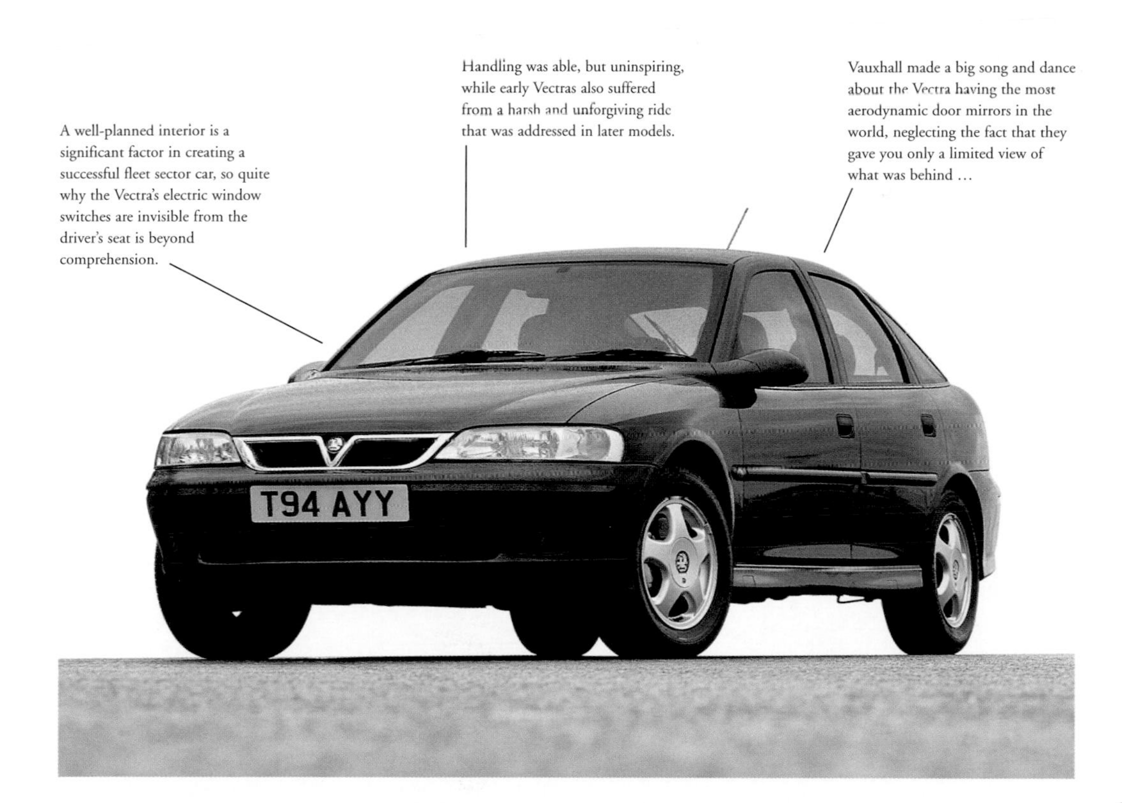

A well-planned interior is a significant factor in creating a successful fleet sector car, so quite why the Vectra's electric window switches are invisible from the driver's seat is beyond comprehension.

Handling was able, but uninspiring, while early Vectras also suffered from a harsh and unforgiving ride that was addressed in later models.

Vauxhall made a big song and dance about the Vectra having the most aerodynamic door mirrors in the world, neglecting the fact that they gave you only a limited view of what was behind …

VOLVO 262C (1977–81)

Every world-famous design studio has a blot in its copybook, and Bertone's lame duck was the Volvo 262C. Built between 1977 and 1981 to special order, the two-door coupé was supposedly the Swedish firm's answer to the BMW 6-Series coupé. It was built along similar lines, using the chassis and basic structure of an executive saloon, but with unique bodywork and a low roofline. But while the BMW was beautifully styled, the 262C was truly awful. It shared the hood, wings and grille of the 264 saloon, but the centre-section was entirely Bertone's own work. That meant the big Volvo got an incongruously low roof and enormous doors, while the vinyl top actually disguised weld marks where the roof of a 264 was chopped to fit.

SPECIFICATIONS

TOP SPEED:	185km/h (115mph)
0–96KM/H (0–60MPH):	12.5secs
ENGINE TYPE:	V6
DISPLACEMENT:	2664cc (163ci)
WEIGHT:	1786kg (3197lb)
MILEAGE:	13.4l/100km (21mpg)

Left: ***Imagine the executive suite of a 1970s office block. Now look at this picture. Can you tell the difference?***

The 262C's cabin was certainly different, but the seats were an incongruous blend of Ikea chic and traditional Chesterfield sofa. Nor pretty.

Bertone's chop top did little for the 262C's already ungainly lines, making the nose and tail appear even more upright than they already were.

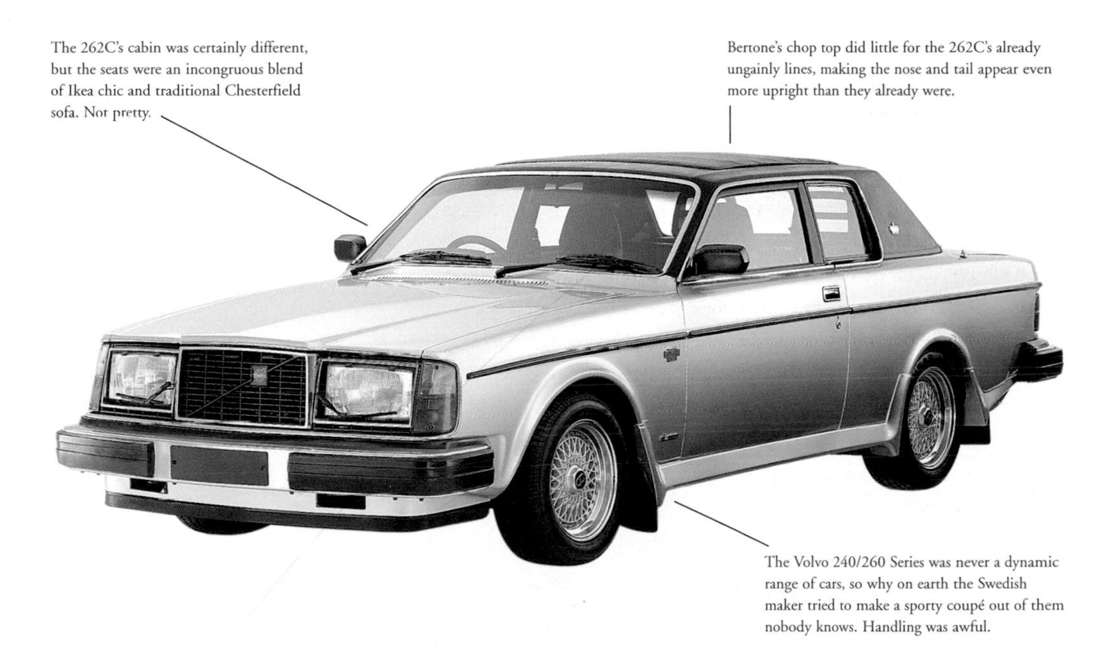

The Volvo 240/260 Series was never a dynamic range of cars, so why on earth the Swedish maker tried to make a sporty coupé out of them nobody knows. Handling was awful.

VOLVO 340 (1975–92)

When is a Volvo not a Volvo? Answer: When it's built in Holland by a lorry manufacturer. The ugly 340 series was originally planned to replace the DAF 66, but, when Volvo took over the Dutch firm in 1975, it launched it as its own. Engines were bought under licence from Renault and were the same units that powered the Renault 5, 9 and 11, meaning they were reliable but unrefined.

The gearbox was mounted above the rear axle, necessitating the use of an extraordinarily long linkage and ruining the gear shift. Handling was stodgy, and rear cart springs were years behind most independently sprung rivals. And yet, awful as it was, the 340 was a tremendous sales success. Especially in the UK market, where it remained in the top 10 bestseller lists right up to its death in 1992, by which time it had become a million-seller.

SPECIFICATIONS

TOP SPEED:	152km/h (94mph)
0–96KM/H (0–60MPH):	15.0secs
ENGINE TYPE:	in-line four
DISPLACEMENT:	1397cc (85ci)
WEIGHT:	983kg (2184lb)
MILEAGE:	7.8l/100km (35mpg)

Left: ***Volvo's reputation for safety was carried over into the 340, and this illustration shows what the crumple zones did in an accident.***

It was launched at a similar time to classics such as the VW Golf, but the 340 didn't follow traditional hatchback styling cues. In fact, it was a very odd shape.

Power came from a range of outdated overhead valve Renault engines, so performance and refinement weren't great. That said, they're impossible to kill.

There was nothing modern about the 340's suspension. Cart springs at the rear and struts at the front made it tail-happy, especially given the rear-drive layout …

LUCAS
LUCAS

FINANCIAL FAILURES

When it comes to a car being a flop, it's fair to say that a car which bankrupts its manufacturer is a pretty spectacular dud. And many of the vehicles in this section did just that – huge development costs, appalling sales records and a complete lack of showroom appeal meant their makers could never recoup the enormous investment required to bring a car to market. The likes of Bricklin, De Lorean, Edsel and Tucker may never see the light of day again, but there will be other ambitious car builders who grind to a halt, stalled by their own attempts to succeed in one of the most cut-throat industries in the world.

Other cars in this section were reasonably good sellers – the Austin 1100, for example, sold over a million in its 11-year life. But British Leyland lost money on every single car it built. In mass production terms, that's a huge loss. And BL was not alone, as Lancia and Chrysler will testify.

Some entries here did not lose money themselves, but acquired such an appalling reputation that the damage they did to their makers' reputations meant the losses suffered later on were immense.

Left: *The AC3000ME was out of date by the time it reached production, and nearly crippled the company that built it.*

AC 3000 ME *(1979–84)*

With the legendary Cobra in its bloodline, the AC 3000 ME should have been a success story. Here, promised the company, was the Cobra's spiritual successor: an affordable two-seater sports car with a lively rear-drive chassis and a powerful Ford engine. The car was first shown at the 1973 London Motor Show as the Diablo, and received rave reviews – it was cited as Britain's first affordable supercar. But things quickly went downhill. It took another six years for the 3000 ME to reach production, by which time the price had leapt from the original proposed £4000 to more than three times that. It was horrifically expensive and complicated to build, the handling was terrible and the performance wasn't even all that good. Oh, and it also suffered gearbox problems. It cost millions to develop and practically crippled the company, and only 82 cars were ever built before AC was forced into receivership.

SPECIFICATIONS

TOP SPEED:	193km/h (120mph)
0–96KM/H (0–60MPH):	8.5secs
ENGINE TYPE:	V6
DISPLACEMENT:	2994cc (182ci)
WEIGHT:	1117kg (2483lb)
MILEAGE:	16.6l/100km (18mpg)

Left: *The marketing campaigns were boastful, but not impressive enough to secure sufficient sales to keep the AC 3000ME in production.*

AC made its own gear linkages to mate to the Ford engine, and this became the model's biggest downfall, with frequent gearbox failure.

Power came from a Ford V6 engine mounted in the middle of the chassis, but, despite this layout, weight distribution was poor and the handling was never great.

The 3000ME used a fibreglass body mounted on a complex sheet steel chassis, the two combining to make it difficult – and expensive – to build.

AUSTIN 1100/1300 *(1963–74)*

When it was new, the 1100/1300 series was cited as a great British achievement. Here was a cleverly packaged, front-wheel-drive car with a practical interior, large luggage area and excellent ride and handling. It became a huge sales success, but, despite the many admirable facets to its design, it soon turned into a nightmare. Build quality was dreadful and rust-proofing was almost nonexistent, meaning rampant corrosion sent many an 1100 or 1300 to an early grave. The car also suffered more than its fair share of mechanical maladies. Yet despite a tattered reputation, British Leyland still managed to shift 1.1 million of them. So why was it such a disaster? Simple – in order to undercut Ford, BL offered the car at a bargain price and lost almost £10 on every example sold. When you look at how many were built, that's a huge sum of money!

SPECIFICATIONS

TOP SPEED:	126km/h (78mph)
0–96KM/H (0–60MPH):	22.2secs
ENGINE TYPE:	in-line four
DISPLACEMENT:	1098cc (67ci)
WEIGHT:	801kg (1780lb)
MILEAGE:	8.0l/100km (35mpg)

Left: ***America was landing men on the moon; Britain was churning out rust-riddled, unreliable saloon cars. And still British Leyland indulged itself in jingoistic marketing.***

Rust was a major enemy of the 1100/1300 series, and found its way into the sills, trunk floors, front wings and wheel arches at an alarming pace.

The 1100/1300 was the work of Mini designer Alec Issigonis, and, like the smaller car, it maximized interior space, with the gearbox being mounted in the sump.

Underneath, the 1100/1300 models used BMC's patented Hydrolastic suspension, which used hydraulic fluid as a springing medium instead of conventional springs.

AUSTIN GIPSY *(1958–68)*

With the immeasurable success of the Land Rover, the Austin Gipsy was hastily rushed into production so that BMC could earn itself part of the large crust that came out of mobilizing Britain's military troops. The Gipsy looked almost identical to a Land Rover, but came with a complex – and expensive – rubber based suspension system. It was effective at first, but reliability and rust problems came to light, and it simply wasn't tough enough to satisfy the demands of armed forces, meaning it failed to recoup its development costs. When Rover (and consequently Land Rover) joined the British Leyland Motor Corporation in 1968, it became surplus to requirements, and was swept under the carpet as an expensive mistake, best forgotten.

SPECIFICATIONS

TOP SPEED:	110km/h (68mph)
0–96KM/H (0–60MPH):	no figures available
ENGINE TYPE:	in-line four
DISPLACEMENT:	2188cc (133ci)
WEIGHT:	1512kg (3360lb)
MILEAGE:	14.1l/100km (20mpg)

Left: *The Gipsy was intended as a rugged, easy-to-fix military vehicle, so what did BMC use in its publicity photographs? Why, an elegant woman, of course ...*

The suspension was made out of rubber, which gave it flexibility and a good ride when new, but, as the rubber aged and perished, the ride and handling became terrible.
Unlike its arch rival, the Land Rover, the Gipsy used steel panels instead of aluminium. That made it prone to rot, especially if used in damp conditions.
The 2.2-litre (134ci) engine was fairly powerful, but was geared for torque rather than performance. That gave the Gipsy immense pulling power, but a low top speed.
AOA 466B

BOND BUG *(1970–74)*

Britain's two biggest manufacturers of fibreglass cars merged in the late 1960s, with Reliant taking over struggling Bond. It saw the move as a way of experimenting with some fairly adventurous designs without putting Reliant's reputation at risk, and the first and only fruit was this – the completely bonkers Bug. Available only in dayglow Orange, the Bug looked like a garishly coloured doorstop on three wheels.

It enjoyed limited success, but most buyers found it far too ludicrous for their tastes and it was dropped after four years and a build run of just over 2000 cars. The Bond name had become a joke associated with the mad Bug, and it was never reintroduced, its reputation irrevocably tarnished. Yet despite killing its maker, the Bug enjoys a remarkable survival rate, with almost half of those built still on the road!

SPECIFICATIONS

TOP SPEED:	78mph (126km/h)
0–96KM/H (0–60MPH):	no figures available
ENGINE TYPE:	in-line four
DISPLACEMENT:	748cc (46ci)
WEIGHT:	622kg (1455lb)
MILEAGE:	7.0l/100km (40mpg)

Left: *You can have any colour as long as it's orange! Henry Ford would have been turning in his grave at the prospect of Bond's unique paint scheme. Plenty of space for those important documents, though ...*

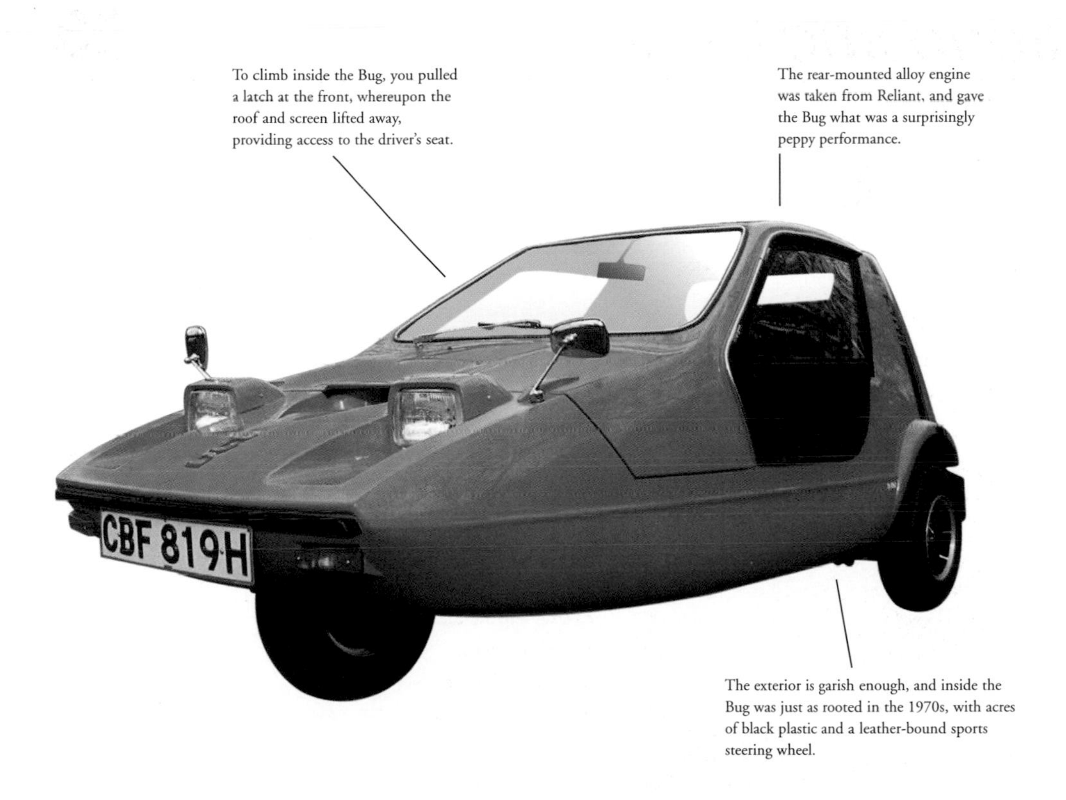

To climb inside the Bug, you pulled a latch at the front, whereupon the roof and screen lifted away, providing access to the driver's seat.

The rear-mounted alloy engine was taken from Reliant, and gave the Bug what was a surprisingly peppy performance.

The exterior is garish enough, and inside the Bug was just as rooted in the 1970s, with acres of black plastic and a leather-bound sports steering wheel.

BRICKLIN SV-1 *(1974–75)*

Canadian entrepreneur Malcolm Bricklin was so confident that his proposed 'Safety Sports Car' would be a huge success that he provided $23 million to build the factory for it, securing the backing of the country's government along the way. The fibreglass-bodied car debuted in 1974 and looked quite stylish, with wedge-shaped lines and gullwing doors – but it was ultimately doomed to failure. Reliability was poor, the AMC-sourced engines were appalling and build quality was dreadful: several owners were trapped inside when the doors packed up. Fewer than 3000 (or a tenth of Malcolm Bricklin's prediction) were built in the first year, and the project collapsed, swallowing all of Bricklin's money and resulting in severe recriminations at government level. One of motoring's biggest ever failures.

Left: *The creator of the SV-1 – the Canadian inventor and businessman, Malcolm Bricklin – whose dream never became a success.*

SPECIFICATIONS

TOP SPEED:	196km/h (122mph)
0–96KM/H (0–60MPH):	8.5secs
ENGINE TYPE:	V8
DISPLACEMENT:	5899cc (360ci)
WEIGHT:	1600kg (3530lb)
MILEAGE:	14.9l/100km (19mpg)

If the battery died, the gullwing doors wouldn't open, meaning the only way to escape was to clamber out of the rear hatch.
The Bricklin was made out of fibreglass, but this wasn't bonded properly, which meant it cracked and warped during climate extremes.
SV stood for 'Safety Vehicle', and that meant it got a distinctly unpleasant lip on the front of the wedge-shaped nose, making it look like it had suffered an accident.
RADIAL T/A

BUGATTI ROYALE (1927–33)

With great success on race tracks and a reputation as a fine automobile maker among Europe's bourgeoisie, Ettore Bugatti was regarded as one of the world's finest car makers. The Royale was his dream – it was supposed to be the ultimate luxury model that would outshine even a Rolls-Royce. Instead, it was an unmitigated disaster. The car was massive and unwieldy to drive, and its own exclusively designed eight-cylinder engine displaced a massive 13-litre capacity. Even the richest buyers saw no point in paying the asking price of 500,000 French francs, which was enough to buy a palace, for a car that only had use as a status symbol.

So, despite great hype surrounding the Bugatti Royale's debut, only six were completed. Of those, only half were ever sold – although the model enjoyed a small amount of fame in 1989, when at US$14 million, it became the most expensive car ever sold at auction.

SPECIFICATIONS

TOP SPEED:	200km/h (124mph)
0–96KM/H (0–60MPH):	no figures available
ENGINE TYPE:	in-line eight
DISPLACEMENT:	12,760cc (778ci)
WEIGHT:	4235kg (9411lb)
MILEAGE:	35.3l/100km (8mpg)

Left: ***Rolls-Royce had its 'Silver Lady' and Jaguar had its 'Leaping Cat'. Bugatti buyers, though, got to admire the famous 'Dancing Elephant' while polishing their chrome.***

The huge 50.8-cm (20-inch) wheels were plated with genuine silver discs, each one costing well over 10,000 French francs to make.

Luxury was definitely the Royale's most distinctive element – the seats were trimmed in ostrich skin, and the dashboard had gold-plated edging.

At 12.7 litres (778ci), the Royale's eight-cylinder in-line engine remains the largest ever fitted to a production model.

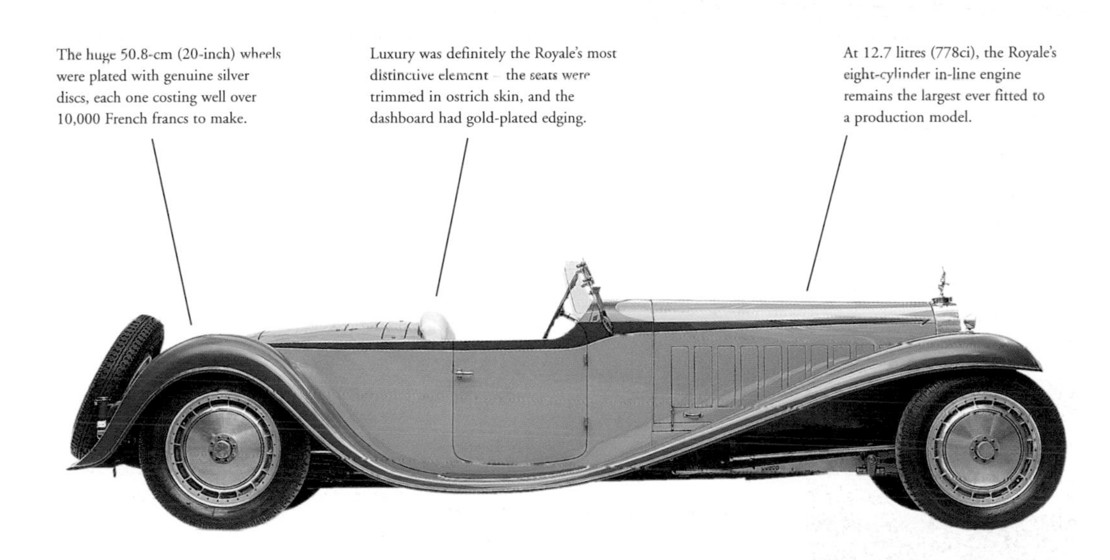

CHRYSLER 180/2-LITRE (1970–80)

An Anglo-French collaboration that came as a result of American giant Chrysler buying Britain's Rootes Group and France's Simca, the 180 was designed to take on Europe's finest luxury cars. In reality, it was an appalling creation. It was dreadful to drive, boring to look at and riddled with corrosion problems, while it also had something of an identity crisis, being known variously as a Talbot, Simca and Chrysler throughout its life. Not surprisingly, the 180 was a flop in showrooms, and residual values were so bad that used car dealers refused to take them in part exchange.

Chrysler lost money on every one sold, but as it had no alternative to offer the market the company kept the car in production for 10 years. A dull and undistinguished offering, with no redeeming features whatsoever.

SPECIFICATIONS

TOP SPEED:	99mph (161km/h)
0–96KM/H (0–60MPH):	13.6secs
ENGINE TYPE:	in-line four
DISPLACEMENT:	1812cc (111ci)
WEIGHT:	1050kg (2334lb)
MILEAGE:	10.0l/100km (28mpg)

Left: ***It's a Chrysler publicity shot, and all of the extras drafted in for the shoot have found something more interesting to look at than the solitary-looking 180. Hardly a surprise ...***

Chrysler really needed to offer something new with the 180, but instead came out with a car that looked like an emaciated Hillman Avenger …

The 180 used a proven 1.8-litre (114ci) engine. Later cars had a 2-litre (122ci) unit, but these were offered only with a power-sapping automatic gearbox.

Shiny black plastics and cheap nylon trim facings weren't what buyers expected from a 'luxury' car, but that's what they got if they bought a 180.

CHRYSLER GAS TURBINE CAR (1963)

The jet engine had such an impact on the aviation industry that Chrysler thought it would be a sure-fire success in the private car market, too. So, in 1963, the American giant built 45 Chrysler Gas Turbine models and sent them out for evaluation to ordinary American families. The car looked futuristic, with lots of round surfaces to emulate the turbine engine's characteristic shape, while inside the vehicle were dials that resembled those of an aircraft cockpit.

It was fascinating as a piece of engineering, but a complete failure as a car. Chrysler's guinea pig families were terrified by the engine's ability to spin to over 44,000 revs, while the handling was unable to cope with the immense performance. Fuel economy was as dreadful as could be expected, and after a year's trial, all but nine cars were recalled and broken for scrap.

SPECIFICATIONS

TOP SPEED:	185km/h (115mph)
0–96KM/H (0–60MPH):	10.0secs
ENGINE TYPE:	gas turbine
DISPLACEMENT:	not applicable
WEIGHT:	1755kg (3900lb)
MILEAGE:	no figures available

Left: ***The Chrysler Gas Turbine car was so nearly a production reality. The American maker sent 50 vehicles out on trial to American families.***

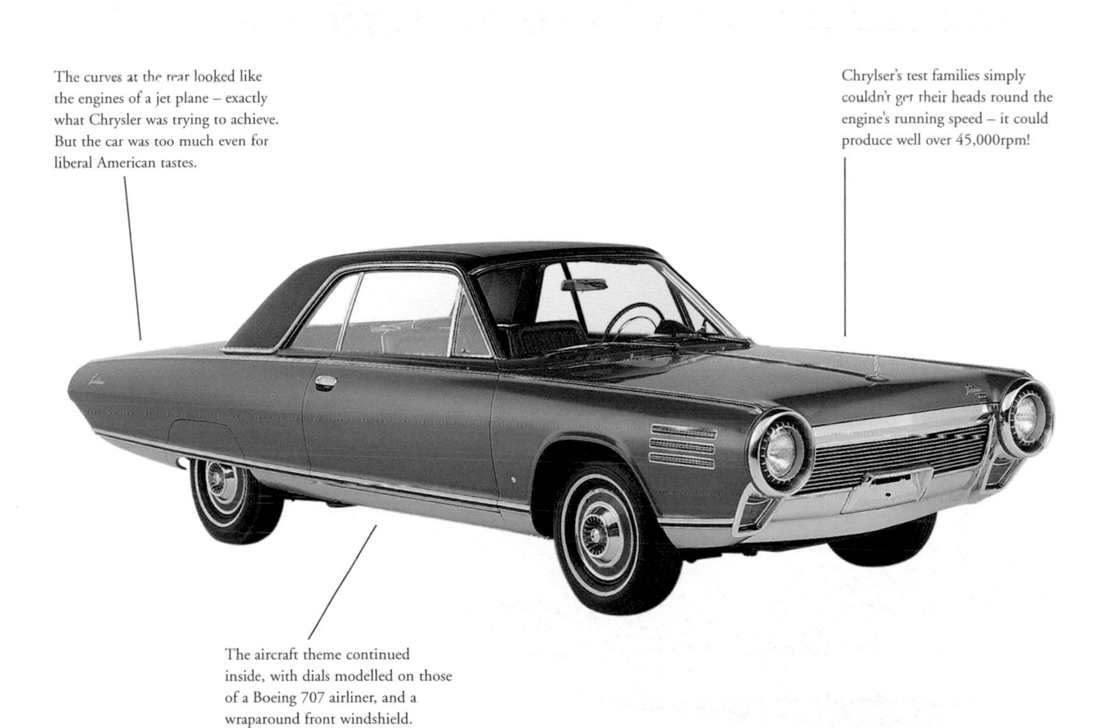

The curves at the rear looked like the engines of a jet plane – exactly what Chrysler was trying to achieve. But the car was too much even for liberal American tastes.

Chrylser's test families simply couldn't get their heads round the engine's running speed – it could produce well over 45,000rpm!

The aircraft theme continued inside, with dials modelled on those of a Boeing 707 airliner, and a wraparound front windshield.

DAF 33/DAFFODIL *(1967–75)*

The Netherlands has never had much of a car industry, but it did have one sales success. That car was the Daffodil, or 33, and more than 300,000 were sold in an eight-year production run. It was a fascinating piece of automotive engineering, offering compact but spacious transport at a sensible price. However, it also had an interesting transmission, made of a series of belt-driven cones attached to the axles and split by a centrifugal clutch. As a result, it was unnecessarily expensive to build, leading to several financial problems for the company that wouldn't be overcome until Swedish maker Volvo took over the helm in 1975, writing off Daf's massive debts.

Perhaps the Daffodil's most intriguing aspect was that, thanks to the transmission, it could go just as quickly in reverse as it could when driven forwards! In truth, though, it was a terrible car which wrecked the company that built it.

SPECIFICATIONS

TOP SPEED:	81km/h (50mph)
0–96KM/H (0–60MPH):	not possible
ENGINE TYPE:	flat-twin
DISPLACEMENT:	590cc (36ci)
WEIGHT:	571kg (1268lb)
MILEAGE:	7.0l/100km (40mpg)

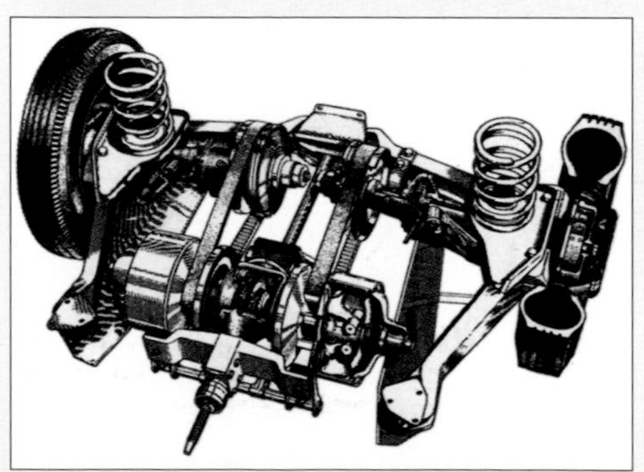

Left: ***This blow-up diagram of the rear transaxle shows how the Daf 33 used a centrifugal clutch and unusual belt-driven transmission.***

Amazingly, the 33's handling was really quite good. It was no larger than a BMC Mini and used the same wheel-at-each corner philosophy.

This was no performance car. It used a 590cc (36ci) flat-twin engine, and struggled to top 80km/h (50mph) at full pelt.

The Daf 33 didn't so much have a gearbox as a direct linkage from the engine to the wheels via a series of rubber belts. It was as fast going backwards as it was going forwards …

DE LOREAN DMC 12 *(1981–82)*

One of the most documented failures in motoring, the De Lorean found fame as the car-based time machine in the *Back to the Future* films. But its screen success wasn't matched in the showrooms, and the car was a total flop. The brainchild of ex-Pontiac man John Z De Lorean, it was built in Northern Ireland, where many hopes were pinned on its success. It was funded by government money, with a chassis by Lotus and styling by Giugiaro. But it was awful. Build quality was dire, the stainless-steel body panels were easily marked and the performance and handling were poor. It bombed, foundering in a trail of debts, corruption and job losses, with accusations of embezzlement, backhanders and bribery among the project's management. And £80 million of British taxpayers' money went down the drain with it.

SPECIFICATIONS

TOP SPEED:	194km/h (121mph)
0–96KM/H (0–60MPH):	10.2secs
ENGINE TYPE:	V6
DISPLACEMENT:	2849cc (174ci)
WEIGHT:	1392kg (3093lb)
MILEAGE:	11.8l/100km (24mpg)

Left: ***The DMC 12 might have been the subject of various untruths and cover-ups, but De Lorean certainly wasn't lying when it claimed the car had 'distinctive looks from any direction'!***

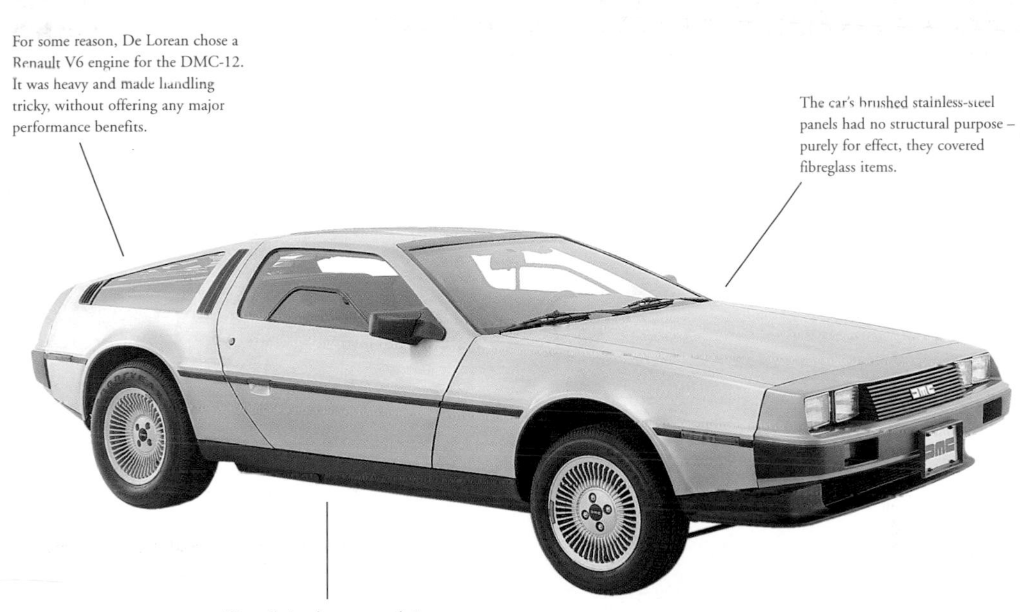

For some reason, De Lorean chose a Renault V6 engine for the DMC-12. It was heavy and made handling tricky, without offering any major performance benefits.

The car's brushed stainless-steel panels had no structural purpose – purely for effect, they covered fibreglass items.

The gullwing doors were a design feature purely for show, and they weren't properly engineered, often trapping passengers inside the car.

EDSEL (1957–60)

Ford blew an astonishing US$300 million on the Edsel project. The aim was to introduce a mid-market car that owners of more basic Fords could aspire to, and the badge was taken from the Christian name of Henry Ford's only son. But the car was ruined by its awkward styling and poor timing. It was launched at a time when many American families were swapping their huge iron 'land yachts' for smaller and more practical cars, and the flashy Edsel had all the hallmarks of cars of the old era. Then there was the unusual 'horse collar' grille, which gave it a somewhat vulgar-looking snout and managed to put many potential buyers off.

The Edsel brand was killed after three years, by which time it had managed to cripple Ford and bankrupt several dealers who had optimistically over-ordered …

SPECIFICATIONS

TOP SPEED:	173km/h (108mph)
0–96KM/H (0–60MPH):	11.8secs
ENGINE TYPE:	V8
DISPLACEMENT:	5915cc (361ci)
WEIGHT:	1747kg (3853lb)
MILEAGE:	15.6l/100km (18mpg)

Left: ***A new kind of popular car: did Edsel understand what 'popular' means?***

Fit and finish was terrible – the rear fin panels sometimes broke off if the car was being driven quickly.

The Edsel's most distinctive styling cue was its vulgar radiator grille, which resembled a horse collar – some critics made far ruder comparisons.

It was billed as a new car, but the Edsel used traditional construction methods and had fairly scary handling.

FORD EXPLORER (1994–2003)

The Explorer is a classic case of a car that in itself wasn't too bad, but which became the victim of circumstances and dire business management. In design terms, it was fine – a huge SUV with plenty of space and good standard equipment. But it was hit by one of the biggest safety scandals in motoring history, resulting in several multimillion-dollar lawsuits. Between them, Ford and Firestone developed special tyres for the Explorer, and these were fitted as standard. But after less than a year, it became apparent that the rubber was a huge problem. Tyres blew out randomly thanks to a design flaw, and the Explorer's top-heavy dynamics meant the car would often roll as a result. After several deaths and serious injuries, along with an ill-fated cover-up operation, the Explorer got new tyres, and the two companies behind the disaster had to compensate the many victims.

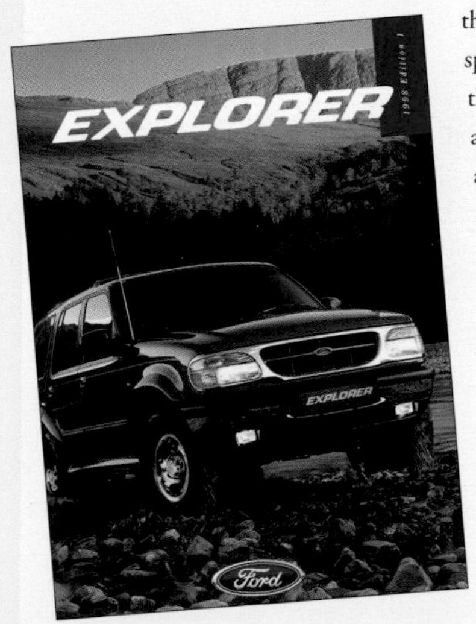

SPECIFICATIONS	
TOP SPEED:	180km/h (112mph)
0–96KM/H (0–60MPH):	9.9secs
ENGINE TYPE:	V6
DISPLACEMENT:	4015cc (245ci)
WEIGHT:	2025kg (4500lb)
MILEAGE:	15.6l/100km (18mpg)

Left: ***An Explorer tackling an off-road course, but most were found bouncing up kerbs outside primary schools and shopping malls.***

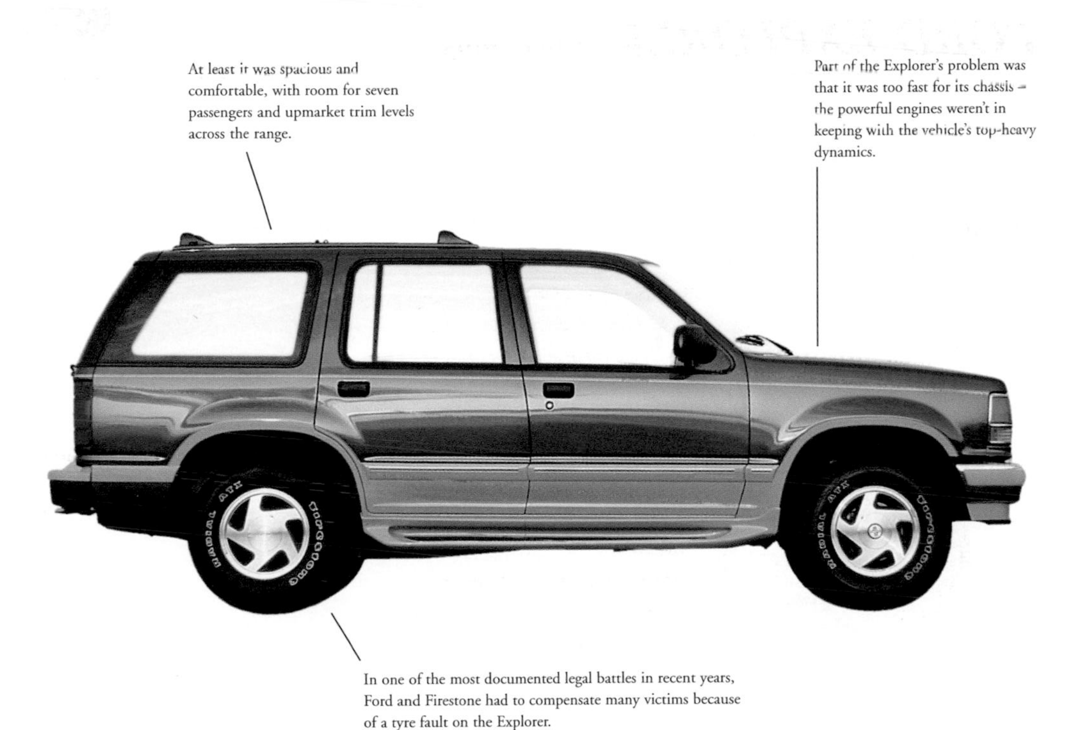

At least it was spacious and comfortable, with room for seven passengers and upmarket trim levels across the range.

Part of the Explorer's problem was that it was too fast for its chassis – the powerful engines weren't in keeping with the vehicle's top-heavy dynamics.

In one of the most documented legal battles in recent years, Ford and Firestone had to compensate many victims because of a tyre fault on the Explorer.

ISO FIDIA *(1967–74)*

After surprising success with its Rivolta coupé and Grifo supercar, Italian domestic appliance manufacturer Iso decided to chart more conventional territory with its next offering. The Fidia was introduced in 1967 as a four-door executive saloon, and was fairly attractive. Power came from a General Motors V8 engine. But the big problem was the cost of buying one: the Fidia was more expensive to develop than Iso had expected, and to recoup the deficit the purchase price kept going up. In the end, it was more expensive to buy than a Rolls-Royce, and for that you got a car that was poorly assembled and prone to rot. By 1974, fewer than 200 cars had been sold and the oil crisis had hit the Italian firm hard. No surprise, then, that it went bankrupt.

SPECIFICATIONS

TOP SPEED:	210km/h (130mph)
0–96KM/H (0–60MPH):	7.2secs
ENGINE TYPE:	V8
DISPLACEMENT:	5359cc (327ci)
WEIGHT:	1288kg (2826lb)
MILEAGE:	18.8l/100km (15mpg)

Left: *As you can see, the Fidia's interior was a picture of Italian elegance. The mystery blonde was an optional extra.*

Despite its bulk, the Fidia was quick, thanks to its Ford-sourced V8 engine. However, the handling was no match for its power output.
Hand-stitched leather and polished wood made the Fidia luxurious, but raised the price beyond affordable levels, seriously affecting sales.
The Fidia may not have been attractive, but the slab-sided bodywork was certainly distinctive. It was designed by Italian styling house Ghia, of Turin.

ISUZU PIAZZA TURBO (1986–90)

After commercial success building lorries and four-wheel-drive vehicles, Japanese firm Isuzu decided to try its hand at a sports car. But with no experience in this area of the market, it turned to smaller companies for assistance. The first to be called in was Italdesign, which offered Isuzu the use of the 'Ace of Spades' concept car seen at the 181 Geneva Motor Show. The concept was developed into a production reality, but to cut costs it used a General Motors platform plucked from the Vauxhall Chevette – a car not renowned for its sporting agility.

The car's handling was atrocious and, to make matters worse, the build quality was also dire. Isuzu turned to Lotus for help in developing the Piazza's handling, but by then it was too late and the damage had been done, costing the company millions. Isuzu never ventured into car building again.

SPECIFICATIONS

TOP SPEED:	205km/h (127mph)
0–96KM/H (0–60MPH):	8.4secs
ENGINE TYPE:	in-line four
DISPLACEMENT:	1996cc (122ci)
WEIGHT:	2810lb (1265kg)
MILEAGE:	28mpg (10.l/100km)

Left: ***In profile, it's easy to see the original lines of Guigiaro's 'Ace of Spades' concept, which formed the basis of the Piazza.***

The cabin seemed to have been thrown together, with no design harmony and a haphazard switchgear arrangement, while the seats were uncomfortable.

Build quality was poor, especially on early cars, which rotted around the rear wheel arches, sills, wings and floorpans.

Choosing the Opel Kadett/Vauxhall Chevette chassis for the Piazza was not an inspired move – it was crude and far too tail-happy. Later models were improved, after Lotus lent Isuzu a hand.

LANCIA BETA *(1972–85)*

It's amazing how the failure of one model can completely destroy a manufacturer's reputation. That's exactly what happened with the Lancia Beta. Touted as the car that would transform Lancia, it did – to a crumbling relic of its former self. The Beta's story is tragic, for in design terms it was excellent – a spacious and stylish car that was entertaining to drive and was available as a saloon, coupé, convertible or estate, offering everything that a model range needed. But with Fiat in control and the resultant need to seriously cut costs, Lancia used cheap bought-in metal, and the quality was atrocious. Betas would start to corrode before they'd even left showrooms, and owners would see their cars practically melt away before their very eyes as rust took a hold. A high-profile media exposé destroyed sales.

SPECIFICATIONS

TOP SPEED:	171km/h (106mph)
0–96KM/H (0–60MPH):	10.8secs
ENGINE TYPE:	in-line four
DISPLACEMENT:	1592cc (92ci)
WEIGHT:	1076kg (2392lb)
MILEAGE:	10.0l/100km (28mpg)

Left: ***According to Lancia, the Beta was 'a choice that leaves you with no alternative'. That's because if you tried to get rid of yours, no dealer would take it in part exchange.***

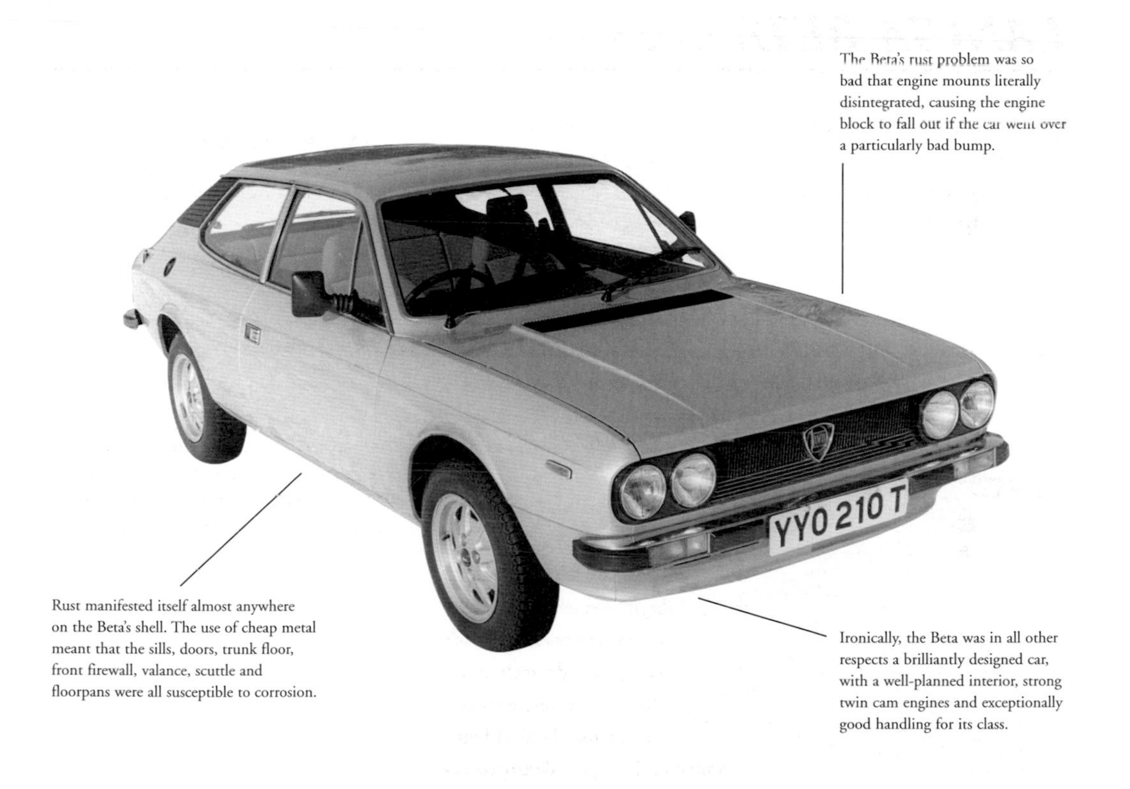

The Beta's rust problem was so bad that engine mounts literally disintegrated, causing the engine block to fall out if the car went over a particularly bad bump.

Rust manifested itself almost anywhere on the Beta's shell. The use of cheap metal meant that the sills, doors, trunk floor, front firewall, valance, scuttle and floorpans were all susceptible to corrosion.

Ironically, the Beta was in all other respects a brilliantly designed car, with a well-planned interior, strong twin cam engines and exceptionally good handling for its class.

LANCIA GAMMA (1976–84)

Another abject lesson in how not to launch a car, the Lancia Gamma had all of the ingredients for sales success, but not the recipe. It was a sharply styled executive car offered as a saloon or a rakish coupé, with a stunning interior and rewarding handling. But it was ruined by dreadful reliability and rampant corrosion, costing Lancia millions in lost sales. Its biggest flaw related to the flat-four 'Boxer' engine. At 2.5 litres, it was huge for a four-cylinder unit, and it was also poorly designed: a flaw in the engineering caused the cambelts to slip off their guides, destroying the cylinder heads and valves, and leaving owners with a hefty engine rebuild bill. Later models were revised and were far better, but the damage had been done and the Gamma never became the sales success it should have been.

SPECIFICATIONS

TOP SPEED:	(203km/h) 125mph
0–96KM/H (0–60MPH):	9.2secs
ENGINE TYPE:	flat-four
DISPLACEMENT:	2484cc (151ci)
WEIGHT:	1373kg (3052lb)
MILEAGE:	12.3l/100km (23mpg)

Left: *Claiming that the Gamma was 'built with all the precision of a Lancia' was not a clever idea really, considering the ongoing Beta rust scandal ...*

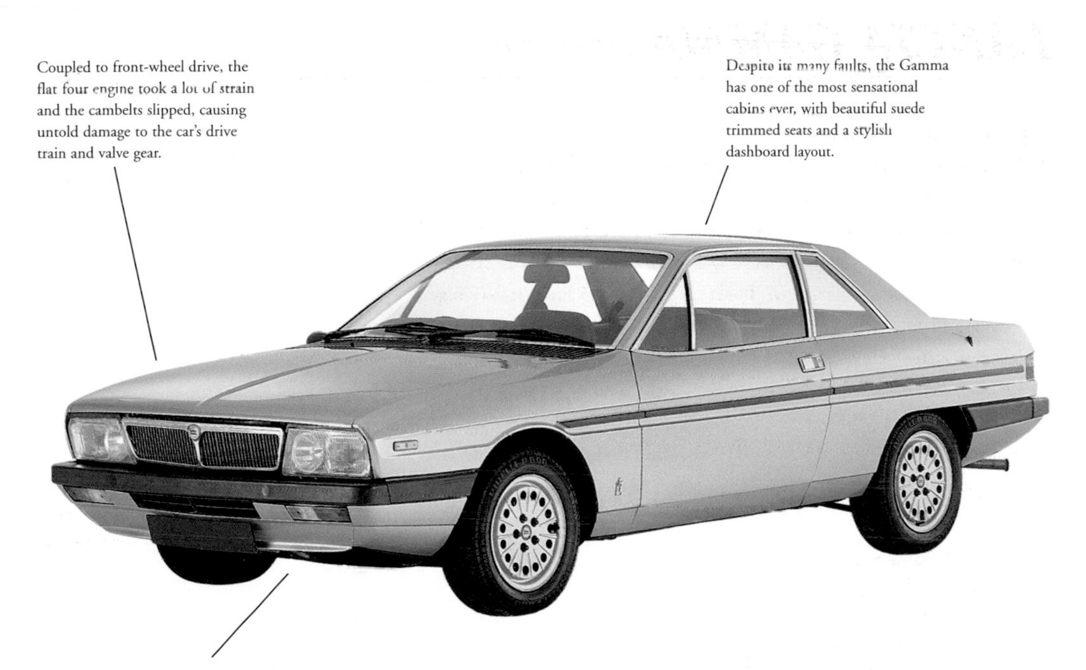

Coupled to front-wheel drive, the flat four engine took a lot of strain and the cambelts slipped, causing untold damage to the car's drive train and valve gear.

Despite its many faults, the Gamma has one of the most sensational cabins ever, with beautiful suede trimmed seats and a stylish dashboard layout.

Gamma buyers could choose from a five-speed manual gearbox or a four-speed auto. The latter was almost a carbon copy of the four-speed AP unit used by British Leyland in the Mini and Allegro.

LEYLAND P76 *(1973–74)*

British Leyland decided to use its links in Australia to create a car exclusively built and sold in the Antipodean region, and the P76 was it. Launched in 1973, the car was meant to compete with market-leading large saloons from Holden and Ford, and came with rear-wheel drive and a high-torque six-cylinder engine. But it soon acquired a reputation as the 'Australian Edsel', thanks to its hideous styling, shocking reliability record and sloppy handling characteristics. Sales were much, much slower than initially anticipated, so to give them a boost Leyland Australia introduced a sporty coupé model, called the Force 7. If anything, this was even worse than the P76, and it was an even greater flop in the sales charts. After two years, the P76 project was canned, with huge financial losses sustained.

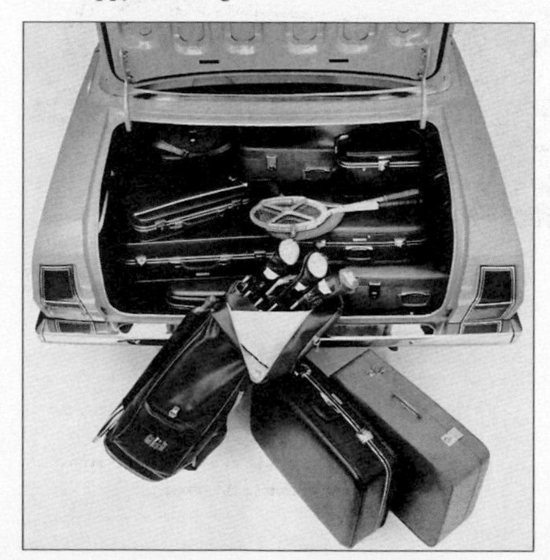

SPECIFICATIONS

TOP SPEED:	168km/h (105mph)
0–96KM/H (0–60MPH):	no figures available
ENGINE TYPE:	in-line six
DISPLACEMENT:	2622cc (160ci)
WEIGHT:	no figures available
MILEAGE:	no figures available

Left: ***In a bid to boost the P76's appeal, Leyland went to great lengths to show buyers how much bigger and better the car was than rivals, but the sales figures proved that few were convinced.***

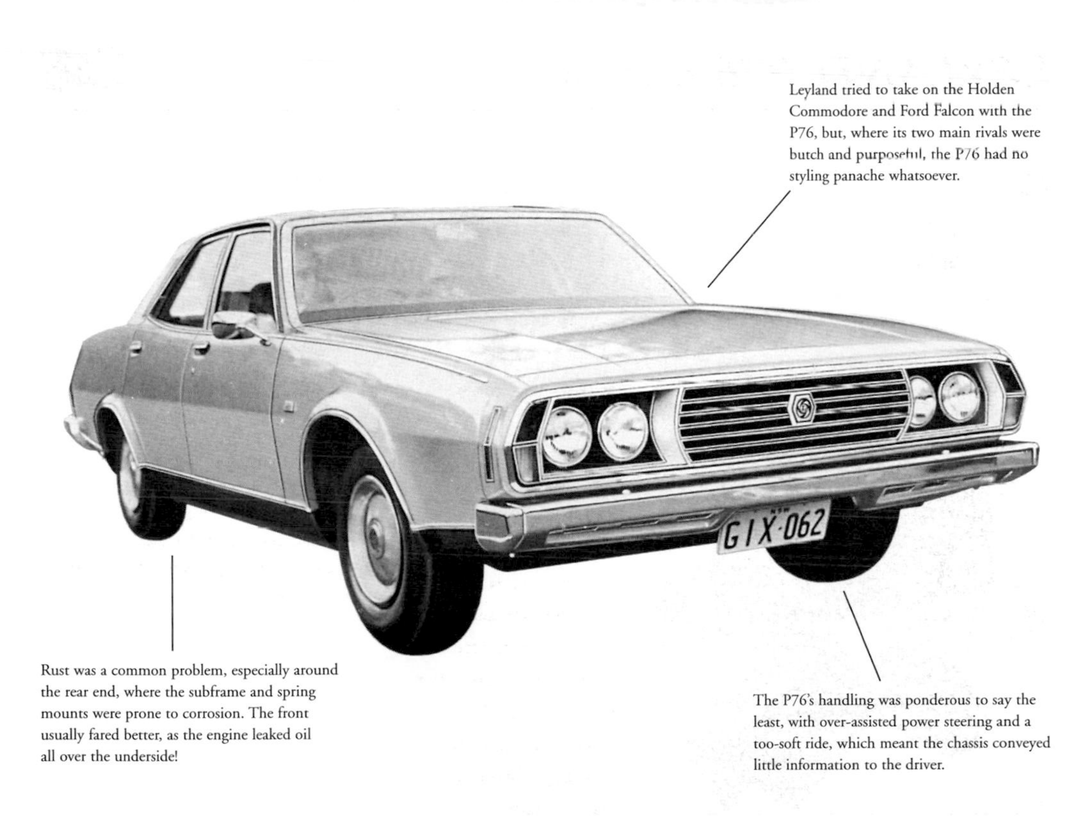

Leyland tried to take on the Holden Commodore and Ford Falcon with the P76, but, where its two main rivals were butch and purposeful, the P76 had no styling panache whatsoever.

Rust was a common problem, especially around the rear end, where the subframe and spring mounts were prone to corrosion. The front usually fared better, as the engine leaked oil all over the underside!

The P76's handling was ponderous to say the least, with over-assisted power steering and a too-soft ride, which meant the chassis conveyed little information to the driver.

MAZDA COSMO 110S (1967–72)

Although the far more famous NSU Ro80 is regarded as the pioneer of rotary engines, the first commercially available use of Felix Wankel's twin rotor technology was actually by Japanese firm Mazda. The fledgling company introduced the Wankel-engined sports car in 1967, claiming it was the most sensational engine technology in the world. In a sense, Mazda was right, but the car wasn't properly developed and problems didn't take long to manifest themselves. Rotor tip seals wore out quickly, and fuel and oil consumption were gross. To add insult to injury, the Cosmo was also extraordinarily rust-prone. Nor was there any escaping the fact that it was hideously ugly, with lines that gave it the appearance of a duck-billed platypus. Just 1176 cars were built before the project was canned, recouping only a fraction of its development costs.

SPECIFICATIONS

TOP SPEED:	169km/h (105mph)
0–96KM/H (0–60MPH):	10.9secs
ENGINE TYPE:	twin-rotary 'Wankel'
DISPLACEMENT:	1964cc (121ci)
WEIGHT:	950kg (2111lb)
MILEAGE:	14.1l/100km (20mpg)

Left: ***The cutaway drawing clearly shows how Mazda installed the Cosmo's rotary engine. Beneath the weird bodywork, the 110S was utterly conventional.***

The Cosmo's 'Wankel' rotary engine offered lively performance and a wonderfully smooth power delivery, which caused excitement at launch, until the many reliability issues started to manifest.

It was an unconventional car with unconventional styling, and the Cosmo's rear deck was almost as long as the hood, with a tiny passenger cell in the middle.

Japanese makers only discovered rust-proofing in the 1970s, and the Cosmo's unusual shape offered lots of flat surfaces and water traps where rust could quickly take hold. If the engine didn't kill it, structural problems would.

MORGAN PLUS FOUR PLUS *(1963–67)*

For the first and only time in its history, Morgan tried to shake off its reputation for making quaint but antiquated cars with the introduction of the fibreglass-bodied Plus Four Plus in 1963. It was certainly up to date, with sleek MGA-style lines, gas-filled suspension dampers and fully functioning window winders, but underneath it was based on the archaic chassis of the standard Plus Four.

To buy a Plus Four Plus, you needed to be rich. It was almost twice the price of the sensational, and infinitely better, Jaguar E-Type, while it was also plagued by reliability problems. Throw in an appalling ride quality, unpredictable handling and a tiny, uncomfortable cabin and it's easy to see why in the end only 26 cars were completed, costing Morgan thousands in ultimately fruitless development bills.

SPECIFICATIONS

TOP SPEED:	169km/h (105mph)
0–96KM/H (0–60MPH):	no figures available
ENGINE TYPE:	in-line four
DISPLACEMENT:	2138cc (130ci)
WEIGHT:	682kg (1516lb)
MILEAGE:	10.9l/100km (26mpg)

Left: ***The Morgan Plus Four Plus had few friends. The advertising men were obviously feeling sorry for it ...***

It might have looked more modern than a standard Morgan, with its fibreglass, streamlined body, but underneath the Plus Four Plus was a traditional wood structure and steel chassis.

There was hardly any room to move inside the Plus Four Plus, and access was difficult because of the low roof. It was simply too impractical for many customers.

The Plus Four Plus drove like a traditional Morgan, but this did not appeal to the new market that the company was chasing. The ride was totally unforgiving, and the chassis far too tail-happy.

NSU RO80 (1967–77)

Europe's annual Car of the Year award goes to the model deemed to be the absolute best introduced in any one year: a car that's at the top of its class in every respect, and which is guaranteed huge sales success. Occasionally, though, the judges get it wrong, and the 1968 jury was guilty as charged. The Ro80 might have been stunningly styled, incredibly comfortable, and fast, refined and enjoyable to drive, but it was also the first mass-produced car to use a rotary engine, and the judges made no allowances for what that might mean. In reality, it meant a car that suffered from colossal fuel consumption and a dreadful lack of reliability, caused by seals collapsing between the rotor tips and the combustion chamber. Sales were terrible, and the model was dropped after 10 years, leaving a trail of debts in its wake. A shame, really, because in most other respects, it deserved its award.

SPECIFICATIONS

TOP SPEED:	177km/h (110mph)
0–96KM/H (0–60MPH):	13.1secs
ENGINE TYPE:	twin rotary 'Wankel'
DISPLACEMENT:	1990cc (120ci)
WEIGHT:	1210kg (2688lb)
MILEAGE:	14.1l/100km (20mpg)

Left: *If only they knew in advance … Britain's* Car *magazine makes a big fuss about the Ro80's election as 1968's Car of the Year.*

Most owners suffered at least one problem with their rotary engine. Rotor tip sealing failure was the most common fault, but excessive oil consumption and poor fuel economy were also common.

In all other respects, this was a ground-breaking car. It had an enormous cabin, semi-automatic transmission and four-wheel disc brakes fitted as standard.

It looked sensational when it first came out and still looks fresh today – the Ro80 would have been an immense success if the original engines had not been shambolic.

OGLE SX1000 *(1961–63)*

The SX1000 was the brainchild of David Ogle, who was a pioneer of fibreglass-bodied cars and the designer of the immensely successful Reliant Scimitar. His idea was to build a small but luxuriously appointed coupé with modern styling, but when it reached fruition the car was actually rather bulbous and ugly.

It was fairly simple to build, though, using the platform of a Mini Van and a stock BMC A-Series engine. But a tragedy doomed the project: road testers had already criticized the SX1000's flimsy construction and twitchy handling, and their point was proven when Ogle himself lost control of his company car in a high-speed bend and was killed outright in the ensuing accident. Unsurprisingly, investors lost all confidence, and the project died with him.

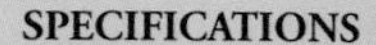

SPECIFICATIONS

TOP SPEED:	177km/h (110mph)
0–96KM/H (0–60MPH):	11secs
ENGINE TYPE:	in-line four
DISPLACEMENT:	1275cc (78ci)
WEIGHT:	701kg (1558lb)
MILEAGE:	8.0l/100km (35mpg)

Left: ***Surely it's no coincidence that the Ogle SX1000 looked vaguely like a squashed Austin Allegro?***

David Ogle was certainly brave with the SX100's styling – it was distinctive, but its bulbous lines and 'frowning' brow did it no visual favours.

The Ogle had a minimalist feeling in keeping with its roots – the cabin was extremely basic, with seating only for two and little in the way of dashboard instruments.

As it was based on a Mini, the Ogle's handling was entertaining, but it could be a real handful at top speed, as it was much lighter than the steel-bodied car on which it was based.

PANTHER SOLO *(1989–90)*

The Solo was Britain's stillborn supercar. Launched to many plaudits at the 1989 Motorfair, the car was stunning to look at and used a complex composite fibreglass and aluminium structure, with power from a turbocharged Ford Cosworth engine. It was great to drive, and was offered at a surprisingly affordable price, but such was the development cost that Panther was on borrowed time by the time the Solo reached showrooms in 1990, and the car was rushed into production to try to reclaim some cash.

The Solo wasn't properly developed, and the four-wheel-drive system was problematic, while the ride was also pretty awful. If it had been engineered properly, the Solo could have been a success, but before Panther had the chance, the receivers stepped in and production was stopped. Only 12 cars had been finished.

SPECIFICATIONS

TOP SPEED:	232km/h (144mph)
0–96KM/H (0–60MPH):	6.8secs
ENGINE TYPE:	in-line four
DISPLACEMENT:	1993cc (121ci)
WEIGHT:	1225kg (2723lb)
MILEAGE:	11.3l/100km (24mpg)

Left: *Panther was keen to show the depth of engineering detail that went into the Solo, but by the time the car reached the showrooms it was a case of too little, too late.*

The mid mounted engine was based on that of the all-conquering Ford Sierra RS Cosworth, but its sharp turbo lag and low gearing were much more suited to rally cars than performance-bred supercars.

Buyers of bespoke sports cars do not want an interior made of other maker's parts, but the Panther's cabin was an awkward mix of Ford and Vauxhall switchgear.

Four-wheel drive should have given assured handling, but it was too complex ever to work properly, and there wasn't enough slip in the differentials to compensate for the steering angle.

RELIANT SCIMITAR SS1 (1984–92)

In concept, the Scimitar SS1 wasn't a bad idea. With the MG Midget and Triumph Spitfire both dead and buried, there was room in the market for a new baby two-seater with a low purchase price and minimal running costs. What let the SS1 down was its execution. For starters, it was ugly. Italian stylist Michelotti started the design, but he died before it was completed, leaving Reliant to finish it off – badly. The wedge-shaped lines were uncompromising, while the build quality was shocking, with gaps in panels and ill-fitting doors. It was originally planned to use the fuel-injected engine from the Ford Escort XR3i, but the hood was too low to shoehorn it in, so in the end it made do with a much less powerful carburettor engine. Reliant lost so much money on the project that it never recovered, and went into receivership.

SPECIFICATIONS

TOP SPEED:	174km/h (108mph)
0–96KM/H (0–60MPH):	11.5secs
ENGINE TYPE:	in-line four
DISPLACEMENT:	1596cc (92ci)
WEIGHT:	873kg (1939lb)
MILEAGE:	10.0l/100km (28mpg)

Left: ***Reliant's 'getaway car' appears to be heading South. Much like the company's profits, then ...***

Oversights do not come much bigger than this: the designers forgot to measure the depth of the proposed Ford fuel-injected engine and it wouldn't fit. The Reliant made do with carburettors until, much later, a fuel-injected Nissan unit was made to fit.

The SS1's pop-up headlights were a common fault – due to rust on the electrical circuitry, they would randomly pack up.

Awkward styling didn't help the SS1's case, but it was the appalling panel fit that ruined it – the doors and trunk lid never closed properly.

ROLLS-ROYCE CAMARGUE (1975–85)

The Camargue was a classic case of a car manufacturer trying to attract a new type of clientele, and failing miserably. The striking two-door coupé was styled for Rolls-Royce by Pininfarina, and was intended to bring a more youthful type of buyer into the company's market. But it didn't work – for starters, the Camargue was the most expensive car on sale at the time, and those who could afford it found its slab-sided lines rather vulgar.

Then there was the build quality, which was nowhere near as good as that of the Silver Shadow, with rust-prone sills and wheel arches. Most buyers opted instead for the more traditional, infinitely better looking Corniche model, leaving the Camargue loitering in showrooms as a costly mistake – one Rolls-Royce never repeated.

SPECIFICATIONS

TOP SPEED:	190km/h (118mph)
0–96KM/H (0–60MPH):	10.0secs
ENGINE TYPE:	V8
DISPLACEMENT:	6750cc (412ci)
WEIGHT:	2328kg (5175lb)
MILEAGE:	18.8secs (15mpg)

Left: *The registration number 1800TU, seen on this brochure cover, has graced Rolls-Royce and Bentley's demonstrator cars since the 1920s ...*

The Camargue was styled by Pininfarina, but this was not one of the Italian styling house's better offerings, with thick slab sides and too much flat surfacing. Traditional Rolls-Royce fans hated it.

It might have looked modern, but under the hood the Camargue had Rolls-Royce's proven and established V8 engine.

Sumptuous surroundings are what every Rolls-Royce owner expects, and the Camargue obliged, with thick leather seats, split-level air conditioning and a polished walnut dash.

STUDEBAKER AVANTI *(1963–64)*

By the time the Avanti appeared in 1962, Studebaker was struggling: it didn't have the factory capacity to compete with more mainstream marques, and it decided that the only way to survive was to offer individual niche products bristling with luxury and technology. But while the Avanti was indeed advanced, with disc brakes and a modern, comfortable cabin, it was a commercial flop that destroyed Studebaker as a car maker for good. The Raymond Loewy-designed body was just too wacky for most tastes, build quality wasn't great and the Avanti was far too expensive for what it was.

Buyers weren't fooled and it was dropped after a year, when the company folded. An almost identical car, the Avanti II, was reintroduced in 1965 by enthusiasts Nate Altman and Leo Newman, and this remained in strictly limited production well into the 1990s.

SPECIFICATIONS

TOP SPEED:	200km/h (125mph)
0–96KM/H (0–60MPH):	11.1secs
ENGINE TYPE:	V8
DISPLACEMENT:	4763cc (291ci)
WEIGHT:	1350kg (3000lb)
MILEAGE:	13.4l/100km (21mpg)

Left: ***The ornate dashboard was one of the Avanti's most elegant styling features. Note the polished aluminium steering wheel boss.***

The Avanti was made out of fibreglass, but the technology was in its infancy, and the first cars were prone to crazing and cracking, especially in hot climates.

The Avanti was certainly distinctive. It was styled by Raymond Loewy, who was responsible for American icons such as the curved Coca-Cola bottle, Lucky Strike cigarette packet and Shell Oil logo.

Getting parts to the Studebaker factory was a problem, as its location in South Bend, Indiana, was miles away from the rest of the US auto industry. Many cars left the factory awaiting parts such as exhaust systems.

Amazingly, the Avanti was one of the first American cars to have disc brakes fitted as standard, one of the few areas in which the US market was well behind Europe.

SUBARU SVX (1991–95)

Before it enjoyed immense rally success, Subaru was renowned as a manufacturer of reliable, tough but ultimately dull four-wheel-drive saloons and estates popular with rural dwellers, but with little showroom appeal elsewhere. The SVX was the firm's second attempt at changing this image, after the ill-fated XT Coupé of the mid-1980s. Launched in 1991, the SVX had four-wheel drive and a Boxer engine, both Subaru trademarks. It was rewarding to drive and had a limpet-like grip, but in styling terms, it was a complete disaster. An amalgam of ideas from Italdesign and Subaru's own designers, the lines had no harmony, and the side detailing, excessive glass and incongruous shut lines were just plain weird. It was also far too expensive. Only a handful were built, reclaiming only a fraction of development costs.

SPECIFICATIONS

TOP SPEED:	232km/h (144mph)
0–96KM/H (0–60MPH):	8.7secs
ENGINE TYPE:	flat-six
DISPLACEMENT:	3319cc (202ci)
WEIGHT:	1602kg (3559lb)
MILEAGE:	11.8l/100km (24mpg)

Left: ***Subaru had an all-wheel-drive heritage to live up to, and the SVX was true to form. This diagram clearly shows the complexity of its engineering.***

The flat-six engine offered plenty of low down grunt and gave the SVX respectable performance, but it came at the expense of fuel consumption, which was especially poor.

The SVX had immense grip, thanks to its four-wheel-drive chassis derived from the Subaru Legacy. If it had not been overpriced and awkwardly styled, it would have been a huge success for its Japanese manufacturer.

Cheap plastics and bland design ruined the interior of the SVX – given the model's somewhat wacky external styling, it was disappointing that the theme was not carried through to the cabin.

TALBOT TAGORA (1981–83)

The Tagora is quite possibly one of the most pointless cars ever built. When Peugeot took over the European arm of Chrysler in 1978, the car was already well into its development stages as an intended replacement for the spectacularly unsuccessful Chrysler 180.

However, instead of doing the most obvious thing and canning the project – which would have been sensible since the 505 Peugeot already had a half-decent executive car – it pressed on with the design and launched the Tagora under the Talbot brand name. Had the Tagora been a good car, Peugeot's choice might have been understandable.

But it wasn't a good car, and not by a long way. The styling was possibly even duller than that of the 180, the handling was atrocious, the interior fell apart and it rusted. It remained in production for three years before Peugeot finally pulled the plug and cut its losses, which were not insubstantial.

SPECIFICATIONS

TOP SPEED:	176km/h (106mph)
0–96KM/H (0–60MPH):	11.3secs
ENGINE TYPE:	in-line four
DISPLACEMENT:	2155cc (131ci)
WEIGHT:	1259kg (2797lb)
MILEAGE:	11.3secs (25mpg)

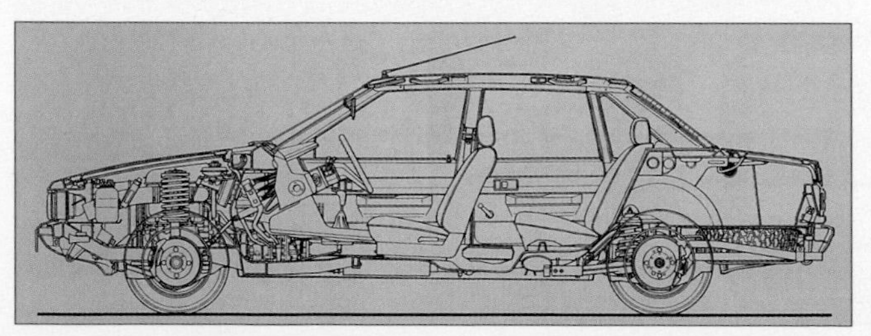

Left: ***Far from showing how advanced the Tagora design was, the technical drawings in the brochure served only to demonstrate just how conventional the new executive saloon's engineering was.***

Handling was uninspiring. The Tagora offered a soggy ride and poor damping.
The executive market demanded a quality finish and high-class materials. The Talbot Tagora offered neither, and was crude and cheap inside
After the lame duck that was the Chrylser 180, the Tagora really needed to be distinctive. Unfortunately, somebody forgot to tell the designers …
TAGORA

TUCKER TORPEDO (1948–49)

It was almost a spectacular success, but the Tucker Torpedo instead became the ultimate failed dream. It was the brainchild of millionaire businessman Preston Tucker, who wanted to introduce the most advanced car ever built. He secured the backing of several wealthy investors to help complete the project, and when it arrived the Torpedo was much lauded. It was aerodynamic, spacious and had a middle headlight that turned with the steering to help the driver see round corners.

But then Tucker was indicted on fraud charges. He was later acquitted, but it was too late and the investors had withdrawn their funds. What should have been one of the finest automobiles ever built became a laughing stock, and just 37 cars were completed before the factory closed down.

SPECIFICATIONS

TOP SPEED:	193km/h (120mph)
0–96KM/H (0–60MPH):	10.1secs
ENGINE TYPE:	flat-six
DISPLACEMENT:	5491cc (335ci)
WEIGHT:	1909kg (4200lb)
MILEAGE:	14.2l/100km (20mpg)

Left: ***The story of the Torpedo was so fascinating that it was made into a film, starring Jeff Bridges as Preston Tucker.***

Power came from a flat-six air cooled engine of 5.5 litres (335ci), which gave the Tucker surprising performance for such a big car. It could cruise at well over 160km/h (100mph).

The Tucker's middle headlight was an inspired piece of design – it turned with the steering to light the way ahead when the driver was negotiating tight bends.

Not only was it a great piece of styling, but the Tucker was also one of the most streamlined shapes ever to grace the roads. If only the project had been given a chance, it could have been an enormous success.

VW K70 *(1970–74)*

Having secured a strong reputation with the air-cooled Beetle and Type 3, Volkswagen tried to capture the conventional market with the K70 – a water-cooled compact saloon car. It acquired the design when it took over struggling NSU in 1969, and continued to develop it, introducing a car that was well made, good to drive and very refined. But it was an unmitigated financial disaster. Volkswagen had built its reputation on interchangeable parts, which made its cars inexpensive to produce.

Nothing on the K70 was shared with other VWs, and its build costs were so high that when it reached showrooms it was significantly more expensive than most of its rivals. That meant the company sold nowhere near as many K70s as it would have liked and lost a fortune, before replacing it with the much more successful Golf in 1974.

SPECIFICATIONS

TOP SPEED:	162km/h (100mph)
0–96KM/H (0–60MPH):	15.6secs
ENGINE TYPE:	in-line four
DISPLACEMENT:	1602cc (92ci)
WEIGHT:	1045kg (2322lb)
MILEAGE:	10.0l/100km (28mpg)

Left: ***The K70's design certainly was advanced – so much so that it cost VW far too much to build and never turned a profit.***

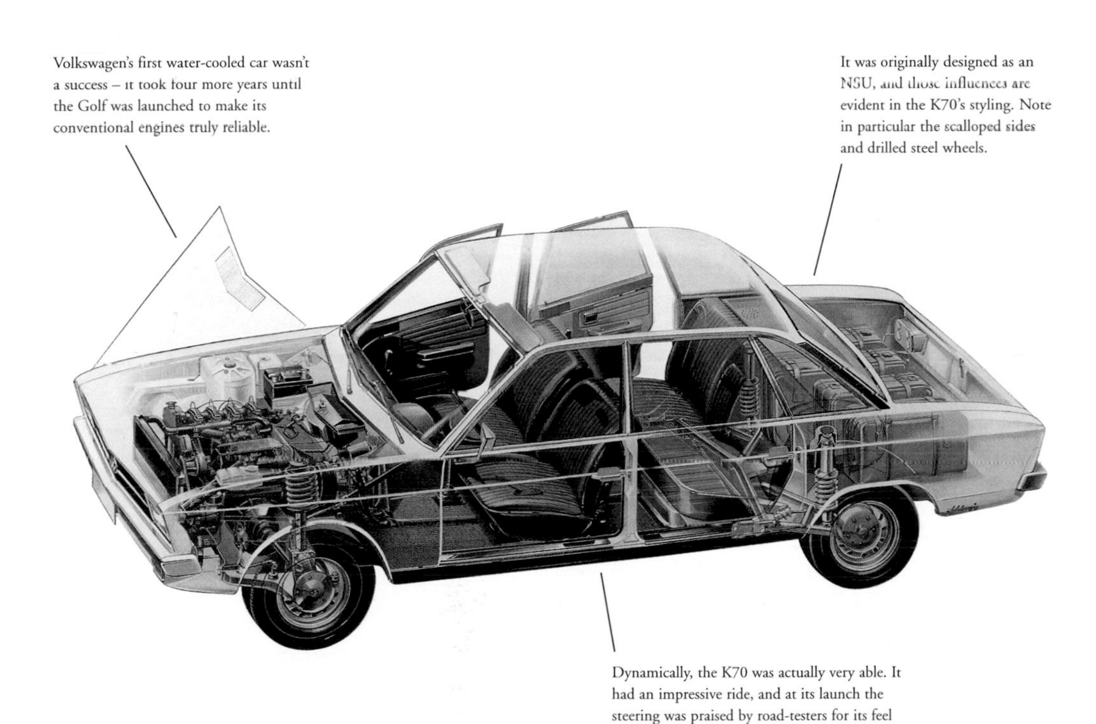

Volkswagen's first water-cooled car wasn't a success – it took four more years until the Golf was launched to make its conventional engines truly reliable.

It was originally designed as an NSU, and those influences are evident in the K70's styling. Note in particular the scalloped sides and drilled steel wheels.

Dynamically, the K70 was actually very able. It had an impressive ride, and at its launch the steering was praised by road-testers for its feel and accuracy.

YUGO SANA (1989–93)

After years of selling rehashed, ancient Fiat designs at bargain prices, Yugoslavian maker Yugo finally introduced its own design in 1989. The Sana looked modern and was even quite a good drive, but the lack of attention to detail in its design and the dire build quality shattered buyers' illusions almost straight away. Given further development, it could have been transformed into an acceptable and generously well equipped budget family car, but by the early 1990s Yugoslavia had started to self-destruct as the civil war gathered pace.

Production stopped, rather abruptly, after the Serbian factory was bombed to pieces and Yugo's insurers refused to cover the damage. Millions of dollars of global investment were wiped out overnight, and the Yugo Sana was never to reappear.

SPECIFICATIONS

TOP SPEED:	160km/h (97mph)
0–96KM/H (0–60MPH):	13.2secs
ENGINE TYPE:	in-line four
DISPLACEMENT:	1372cc (84ci)
WEIGHT:	903kg (2007lb)
MILEAGE:	8.0l/100km (35mpg)

Left: ***The Sana was the first all-new Yugo to debut in 20 years. Before it, the company could only offer cheap and basic 45 (left).***

Italdesign styled the Sana – and it showed. The bodyshell shared styling cues with the Fiat Tipo and Strada/Ritmo, both of which came from the same design house.

If only Yugo had applied more attention to detail, the Sana could have been a greater success, but the cabin was shoddily finished and the plastics were awful.

True to form, Yugo used as many old Fiat bits as possible. The platform came from the Regatta, and the engine, brakes and suspension were cribbed directly from the Uno supermini.

IRAN

MISPLACED MARQUES

The concept of 'badge engineering' has been around since the late 1950s, when the British Motor Corporation took six versions of the same car and put different names on each one, in a bid to cash in on buyers' brand loyalty. It became a global trend, with several other manufacturers following suit. There was a point when buying a Chrysler, Dodge or Plymouth, or a Chevrolet, Pontiac or Buick, got you essentially the same car with a different name on the front. Some of the cars in the following pages show badge engineering taken to shameful extremes.

Here, you will find such cynical naming exercises as a Nissan Cherry with an Alfa Romeo shield on its grille, a downmarket Kia with Ford's blue oval on its nose, and possibly the worst ever misinterpretation of MG's revered octagon, when Austin-Rover saw fit to glue it onto an especially nasty Maestro. Other entries are simply old, outdated models, given new names and marketed in different countries to try to eke out the last ounces of profitability from ageing tooling.

Left: *Many marques live on long after they've been forgotten in their domestic markets. The Paykan is Iran's take on the Hillman Hunter.*

ALFA ROMEO ARNA (1981–85)

Alfa Romeo is one of the finest names in motoring history. Steeped in romantic tradition, the Italian firm has produced some of the most beautiful cars of all time. Unfortunately, it also produced this. The Arna is perhaps the biggest blot on the company's copybook, and was a classic case of desperate times meaning desperate measures. With huge debts and a reputation ruined by the Alfasud, which crumbled at the first sign of rain, the Arna was a last-ditch attempt to win back customers.

Alfa kept the 'Sud's Boxer engine and decided the best place to house it would be the bodyshell of a Nissan Cherry, assuming that the Japanese firm's reputation for reliability would win favour. Sadly, it forgot about the Cherry's awful, dull styling, and, as the engine and electrics still came from Alfa, the car broke down with alarming regularity. It was also sold as the Nissan Cherry Europe, which was even worse.

SPECIFICATIONS

TOP SPEED:	158km/h (98mph)
0–96KM/H (0–60MPH):	13.1secs
ENGINE TYPE:	flat-four
DISPLACEMENT:	1350cc (82ci)
WEIGHT:	843kg (1874lb)
MILEAGE:	8.8l/100km (32mpg)

***Left:** It's easy to spot an Arna thanks to its ornate Alfa Romeo shield, which sits incongruously with the awful plastic Nissan grille. Cherry Europes have their own badging, and most came fitted with a shield marking 50 years of Nissan, in 1985.*

The Arna's cabin was no more exciting than the dull bodywork. The dashboard was standard Nissan Cherry fare, made from cheap-feeling plastic, the only saving grace being a sporty Alfa rev counter.

Power came from Alfa's proven Boxer engine, which at least had some sporting character. Italian buyers disapproved, though, believing that an Alfa engine should only ever appear in an Alfa car, and many of the cars were stripped to restore rotten Alfasuds ...

Part of the reason Alfa teamed up with Nissan was to avoid the rot problem that had plagued the marque throughout the 1970s. But it turned out that the Cherry bodyshell was little better than Alfa's own, and rust problems were still rife, especially around the doors and wheel arches.

The back of an Alfa Romeo Arna: something many owners were glad to see after they'd owned one for a couple of years ...

AMC ALLIANCE/ENCORE (1984–89)

With their dubious styling, shocking reliability and terrible build, cars such as the Pacer and Gremlin had made the American Motor Corporation a laughing stock. To try to redress the balance, AMC asked its biggest shareholder, Renault of France, if it could help with a new design. The French maker duly obliged, and supplied AMC with the body pressings and running gear from the utterly unspectacular Renault 9 and 11 models, which were then built under licence in Wisconsin, complete with hideous federal safety fenders for good measure.

AMC were so confident about the new cars that they even modified the presses to add two of their own designs – a two-door sedan and a soft-top. Needless to say, the entire range was a complete flop – the cars were poorly made, and also too small and slow for American tastes.

SPECIFICATIONS

TOP SPEED:	153km/h (92mph)
0–96KM/H (0–60MPH):	13.2secs
ENGINE TYPE:	in-line four
DISPLACEMENT:	1397cc (85ci)
WEIGHT:	803kg (1786lb)
MILEAGE:	8.3l/100km (34mpg)

Left: *Quite what the advertisers were thinking by plastering the word 'impossible' on the sales literature for the Alliance is beyond us. Perhaps they were talking about reaching their sales targets ...*

Conservative European tastes didn't fit the American market, so AMC embellished the Alliance and Encore with a few inimitable alterations, such as chrome wheel trims and whitewall tyres.

AMC tried a couple of its own versions of the cars, using the Renault 9 body as a basis. The two-door coupé and cabriolet models used the doors from a Renault 11 and a specially strengthened sill section to make up for the loss of the original car's B-pillar.

Renault's 1.4-litre (85ci) and 1.7-litre (104ci) engines were already dated when they appeared in the 9 and 11, so as well as lacking power they were not particularly economical. As high-revving, noisy four-cylinder units, they failed to find favour with American buyers.

In Europe, the Renault 9 and 11 were inoffensive, if a little bland. Then AMC got its hands on them, and they sprouted huge fenders and weird-looking lights, which did little to enhance their appearance.

CHEVROLET CHEVETTE (1974–82)

The mid-1970s fuel crises took American makers by surprise, leaving the larger organizations turning to their European arms for assistance. For Chevrolet, help came in the way of the Chevette, built under licence from UK badge Vauxhall. The British Chevette was ugly enough, but what Chevrolet did to it was criminal, adorning it with whitewall tyres, wood panelling and the nose of the derided Chevy Vega.

Inside, it was wall-to-wall shiny plastic, in your choice of beige, black or embarrassingly lurid red. The car's small pushrod four-cylinder engines just might have kept US buyers happy, were it not for the fact that the Chevette suffered from rapidly encroaching body rot, electrical maladies and a tendency to spin out of control if cornered at any speed. There are few survivors.

SPECIFICATIONS

TOP SPEED:	147km/h (91mph)
0–96KM/H (0–60MPH):	14.5secs
ENGINE TYPE:	in-line four
DISPLACEMENT:	1256cc (76.6ci)
WEIGHT:	846kg (1879lb)
MILEAGE:	34mpg (8.3l/100km)

Left: *'It'll drive you happy,' said Chevrolet. We take it that 'happy' was being used as a euphemism for 'mad' or 'crazy'.*

With a short wheelbase and rear-wheel drive, the Chevette could be twitchy in the wet, while its live rear axle was crudely engineered, making it difficult to pull back into line if the back stepped out.
Oh dear! The fake wood panelling that adorned many a Chevette gave it the look of an upmarket station wagon, or so said Chevrolet. In reality, it served only to make a daft car look even dafter.
Americans liked their cars to be big on the outside, and equally capacious once you climbed in the door. The Chevette claimed to be a genuine five-seater, but space in the rear was so cramped it could house only two in comfort.
The Chevette's nose was modified to distinguish it from European models, losing the original car's sleek profile and replacing it with the front of a Chevy Nova. The excess chrome was completely out of keeping with the car's compact character.
1976

CHRYSLER AVENGER *(1977–82)*

Originally, the Avenger wore a Hillman badge and was a worthy if not exactly inspiring family car. Build quality was good, performance was adequate and the styling was really quite pretty, with its unique 'hockey stick' tail-lights and swooping D-pillars. Then Chrysler's influence took over.

The corporate badge replaced Hillman, while the nose and rear were changed, with excess chrome, hideous horizontal rear lights and a Pentastar logo on almost every panel. Cheaper metal was used, causing the Avenger to suffer from sudden and rather disturbing rust problems, while the interior was given an inadvisable makeover, resulting in multicoloured nylon seat facings and a dashboard coated in black PVC. Falling sales resulted in Chrysler Europe giving up and selling out to Peugeot-Talbot, which put the Avenger out of its misery less than two years later.

SPECIFICATIONS

TOP SPEED:	134k/mh (84mph)
0–96KM/H (0–60MPH):	19.8secs
ENGINE TYPE:	in-line four
DISPLACEMENT:	1295cc (79ci)
WEIGHT:	850kg (1889lb)
MILEAGE:	8.5l/100km (32mpg)

Left: ***Just look at that interior. Quite who Chrysler was trying to fool when it claimed the Avenger was better than its Hillman predecessor is anyone's guess, and the cabin was definitely no improvement ...***

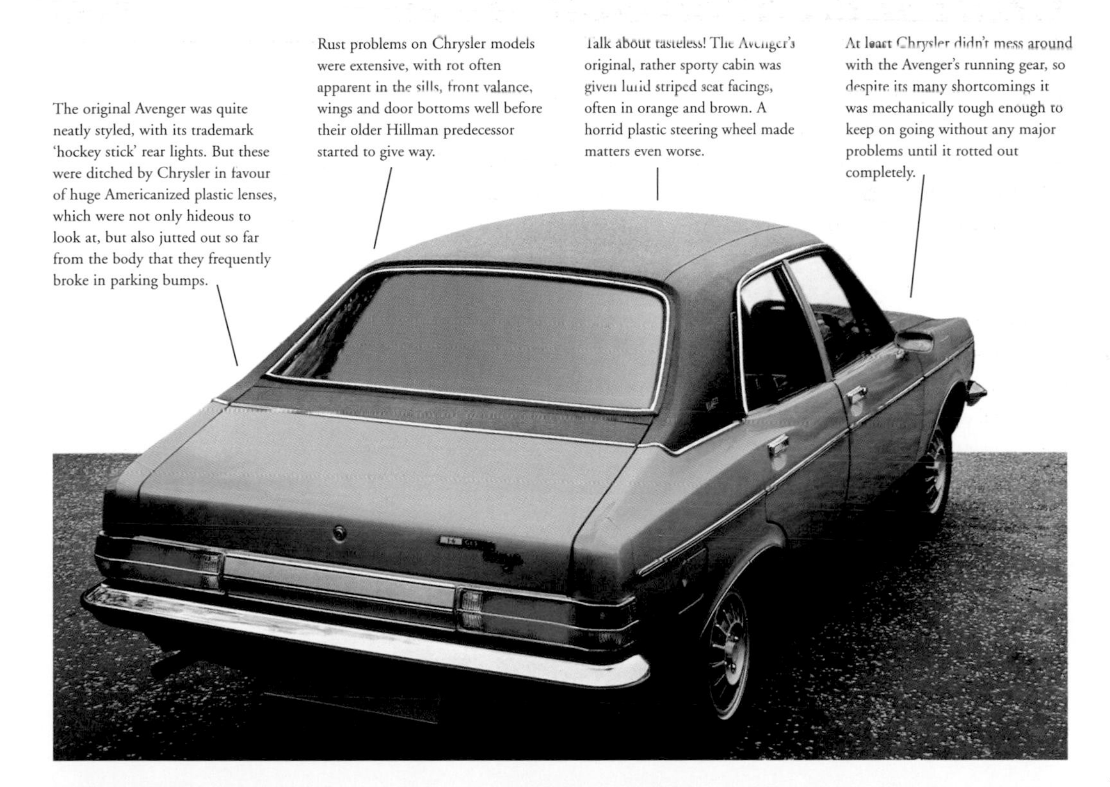

The original Avenger was quite neatly styled, with its trademark 'hockey stick' rear lights. But these were ditched by Chrysler in favour of huge Americanized plastic lenses, which were not only hideous to look at, but also jutted out so far from the body that they frequently broke in parking bumps.

Rust problems on Chrysler models were extensive, with rot often apparent in the sills, front valance, wings and door bottoms well before their older Hillman predecessor started to give way.

Talk about tasteless! The Avenger's original, rather sporty cabin was given lurid striped seat facings, often in orange and brown. A horrid plastic steering wheel made matters even worse.

At least Chrysler didn't mess around with the Avenger's running gear, so despite its many shortcomings it was mechanically tough enough to keep on going without any major problems until it rotted out completely.

CHRYSLER TC MASERATI (1986)

Chrysler was confident that its Le Baron convertible was a fine motor car: a stylish, well-built soft-top that had immense showroom appeal. But the company's name wasn't synonymous with glamorous models, so it turned to the style centre of the world – Italy – for help. The chosen benefactor of a large sum of Chrysler money was struggling Maserati, a company with a fine tradition for making cars with great engines and attractive styling.

Sadly, Maserati was going through a barren period of its own at the time, and build quality was terrible. When Chrysler threw it a lifeline by asking it to supply cylinder heads for the TC, Maserati leapt at the opportunity, and hastily built a batch of Cosworth-tuned powerplants for shipment to America. Over the next few months, almost every single one of them suffered a major mechanical failure, and the TC's reputation was sullied from the start. All-in-all, a disastrous tie-up.

Left: ***The Chrysler company claimed the TC was engineered for dependability, power and performance. Really?***

SPECIFICATIONS

TOP SPEED:	210km/h (130mph)
0–96KM/H (0–60MPH):	8.8secs
ENGINE TYPE:	in-line four
DISPLACEMENT:	2211cc (135ci)
WEIGHT:	1364kg (3033lb)
MILEAGE:	12.8l/100km (22mpg)

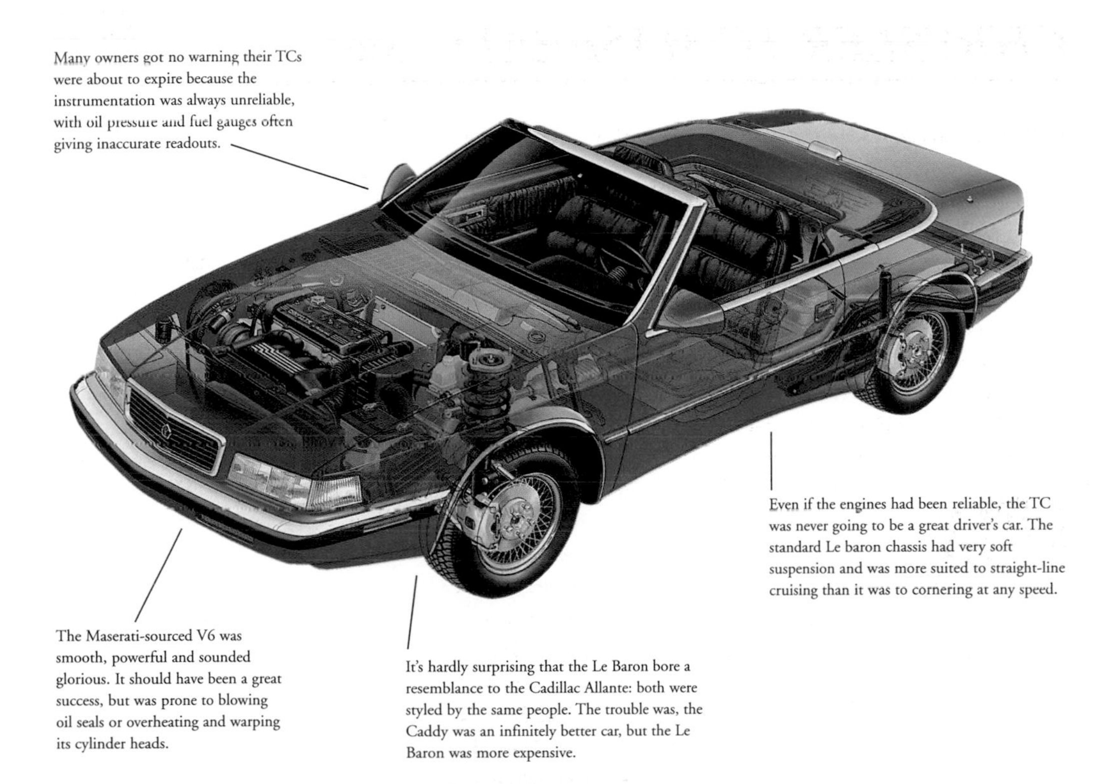
Many owners got no warning their TCs were about to expire because the instrumentation was always unreliable, with oil pressure and fuel gauges often giving inaccurate readouts.
Even if the engines had been reliable, the TC was never going to be a great driver's car. The standard Le baron chassis had very soft suspension and was more suited to straight-line cruising than it was to cornering at any speed.
The Maserati-sourced V6 was smooth, powerful and sounded glorious. It should have been a great success, but was prone to blowing oil seals or overheating and warping its cylinder heads.
It's hardly surprising that the Le Baron bore a resemblance to the Cadillac Allante: both were styled by the same people. The trouble was, the Caddy was an infinitely better car, but the Le Baron was more expensive.

DACIA DENEM *(1983–84)*

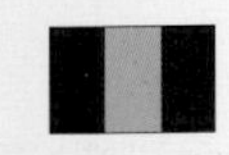

Romanian maker Dacia had been building its own version of the old Renault 12 under licence for years and, to be fair, it was the perfect car for what was then an impoverished Communist dictatorship. Cheap, tough and easy to maintain, it was ideal for a model that wasn't so much sold as allocated to people who had been on the waiting list for almost a decade.

The problems came when Dacia thought it could crack Western Europe, where the car was introduced in 1983 as the Denem. It targeted Britain as its main market, and paraded saloon and estate models at the 1983 Motorfair in the hope of drumming up sales. But it was based on the Renault 12, which was long since defunct, having disappeared from UK showrooms seven years earlier. But buyers weren't about to have the wool pulled over their eyes …

SPECIFICATIONS

TOP SPEED:	143km/h (89mph)
0–96KM/H (0–60MPH):	16.5secs
ENGINE TYPE:	in-line four
DISPLACEMENT:	1289cc (79ci)
WEIGHT:	873kg (1940lb)
MILEAGE:	10.0l/100km (28mpg)

Left: ***Apparently, the Dacia Denem made choosing easier. Well, you did know which car to rule out of your test drive choices.***

Luxury wasn't an option with the Denem. If you went for the top model, you got cloth seat inserts and a temperature gauge, but basic cars had hard-wearing plastic trim, a speedometer and a fuel gauge – and not much else.

Dacia offered the Denem with a 1.4 litre (85ci) Renault pushrod engine, which was tough enough to last for several thousand miles. Sadly, each one was accompanied by a gruff exhaust note and clattery tappets.

The Denem certainly wasn't a car for driving quickly. The crude cart-sprung chassis was built for surviving rough roads, and if you tried to corner at any speed, the inside front wheel lifted in the air and the car lost all its grip.

The Renault 12's styling was ungainly enough, but the Denem was even worse, with a hideous plastic nose cone grafted on the front, incorporating twin headlights for a supposedly 'sporty' look.

DAEWOO RACER/NEXIA *(1995–1999)*

Daewoo was one of many South Korean car makers that sprung to prominence in the late 1980s and early 1990s. The motor manufacturer was part of a massive industrial conglomerate that made everything from ships to toasters, and, much like the rest of its products, the Nexia – or Racer, depending on which market the car was sold in – was built to serve a purpose rather than inspire. It was cheap to build, as it was based on the mid-1980s Vauxhall Astra/Opel Kadett, and the car's new nose did little to disguise the fact.

The Astra was quite good when new, but the Nexia didn't debut until 1995, by which time it was a crude basic design that offered nothing other than value for money. In Europe, Daewoo tried to sell the car directly to buyers, not bothering with dealerships, but its success was more limited than expected, meaning many examples sat unsold on airfields, eventually being broken up for spares. A tedious, uninspiring creation.

SPECIFICATIONS

TOP SPEED:	162k/mh (100mph)
0–96KM/H (0–60MPH):	11.4secs
ENGINE TYPE:	in-line four
DISPLACEMENT:	1498cc (91ci)
WEIGHT:	970kg (2156lb)
MILEAGE:	7.2l/100km (39mpg)

Left: ***The Daewoo badge became so tainted in European markets that it was changed to Chevrolet in 2004.***

Inside, the car was stuffed with Vauxhall-Opel influences, including the same dashboard moulding and seats as the Astra, along with its dated 1980s-style switchgear and window buttons.

Daewoo tried to disguise the Nexia's origins with a hideous plastic grille and an odd-looking vertical badge. A later facelift made the front end even more angular, but the fit and finish were awful.

Spot the old Vauxhall Astra lurking inside the new Daewoo body! The doors, glass and rear quarter panels were exactly the same as the GM hatchback that had debuted more than a decade earlier – and looked far better.

Guess what? Yes, it's GM old stock again. Although the Nexia's 1.5-litre (92ci) engine was never offered in the Astra or its derivatives, it was an old General Motors unit that had previously seen service in compact Holden models in Australia.

FORD ASPIRE *(1995–2002)*

Quite who would aspire to owning one of these is a mystery. Ford's budget compact model, launched in 1995, has to be one of motoring's most obvious misnomers, with awkward, bulbous styling, appalling quality and a cheap and nasty plastic cabin. The car was built in conjunction with Korean maker Kia, who sold it in developing markets as the Avella, with a surprising degree of success.

However, in America, the Aspire wasn't especially successful. Buyers weren't fooled by the name, nor were they wooed by the car's prominent understeering, obstructive gearshift, spongy brakes or awful driving position. Bits fell off with alarming regularity, and, if that wasn't enough, the Aspire also rusted quite badly – an unusual trait for a 1990s model. The Aspire might have been cheap, but those who bought one soon found out why.

SPECIFICATIONS

TOP SPEED:	171k/mh (105mph)
0–96KM/H (0–60MPH):	11.9secs
ENGINE TYPE:	in-line four
DISPLACEMENT:	1498cc (91ci)
WEIGHT:	913kg (2029lb)
MILEAGE:	8.3l/100km (34mpg)

Left: ***The Ford Aspire was built by Kia. Perhaps the wheeltrim was the only genuine Ford part on this awful car.***

The Aspire wasn't exclusively a Ford. It was also sold as a Kia and a Mazda in some markets, and managed to top the bestseller charts in at least one country. Thanks to a large order from Avis Rent-a-Car, it was the best-selling new model in Malta for 1997!

The Aspire housed a high-level brake light in the metalwork of its tailgate, but it was badly designed, and water trapped in the lens would trickle out, causing the trunk lid to rust.

Changing gear in the Aspire was a truly demoralizing experience. The lever felt like it was mounted in a sponge, and it was impossible find the gears without stirring the lever round in an animated circular motion between each change.

Built on one of the most uninspiring chassis ever constructed, the Aspire was prone to understeer at relatively low speeds and had a horrible, choppy ride, while the steering itself was dreadful.

FSO 125P *(1978–1991)*

The Polish-built 125P was originally sold in Western Europe as the Polski-Fiat because the body was that of the 1965 Fiat 125, built under licence in Warsaw. But build quality was so bad that, by the mid-1980s, Fiat asked FSO to remove all reference to its name from the badging, fearing that the 125P's terrible reputation would reflect badly on its own.

So what exactly was wrong with the 125P? After all, the Fiat 125 was regarded as a fine car when it first came out. First, the 125P was two decades behind the times, and was completely undeveloped from the old Fiat design. Its archaic structure could be ignored because of its price: it was the cheapest model on the market in almost every country it was sold. Reliability was shocking. Cheap carburettors used to break up, sending slivers of metal into the fuel supply, the rear springs collapsed for no reason, and the body and chassis were usually rustier than a shipwreck.

SPECIFICATIONS

TOP SPEED:	145km/h (93mph)
0–96KM/H (0–60MPH):	14.4secs
ENGINE TYPE:	in-line four
DISPLACEMENT:	1481cc (90ci)
WEIGHT:	1028kg (2285lb)
MILEAGE:	10.0l/100km (28mpg)

Left: ***The best 125P was the station wagon – it was cheap and had a huge load bay, but otherwise it was as shocking and shoddy as the four-door model.***

Such was the 125P's corrosion problem that if you put something heavy in the trunk of a particularly tired example, it would fall straight through to the floor – taking the gasoline tank with it!

In profile, the 125P was actually quite stylish, thanks to its Italian ex-Fiat styling. But the car was so bad that Fiat quickly withdrew its name from the Polish project for fear of tarnishing its reputation, which was already blighted by rust problems of its own.

If you saw a model with half of the rear wheel covered by the wheel arch, then you knew that the rear spring hangers had gone and that the rear axle was beginning to disappear into the bodywork as the floor around it started to rot. A terminal fault.

The 125P was designed to be easy for a home mechanic to repair – which was just as well because the engines were thrown together quickly and regularly suffered from fuel supply and coolant loss problems.

HINDUSTAN AMBASSADOR (1966–present)

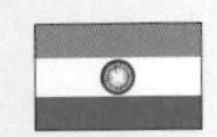

Look at the picture, and you'll find it hard to believe that this car is still on sale as a new model. But the Indian-built Hindustan Ambassador, or 'Amby' as it's affectionately known by its many devout followers, remains one of India's best-selling models.

The car is not so much based on the 1954 Morris Oxford as it is a carbon copy of the car, despite being more than half a century old. In that time, you'd expect something in the way of development to take place, but short of wider fenders to satisfy crash regulations and a catalytic converter to comply with modern vehicle construction regulations, the Ambassador remains very much an undeveloped relic, right down to the lead-heavy steering and brakes that have never heard the phrase 'emergency stop'. With over a million Ambys on India's roads, it's hardly surprising that it has one of the highest road death rates in the world. Yet the Amby still sells in droves, and there's even a diesel version for ultimate purgatory.

SPECIFICATIONS

TOP SPEED:	140km/h (87mph)
0–96KM/H (0–60MPH):	no figures available
ENGINE TYPE:	in-line four
DISPLACEMENT:	1818cc (110ci)
WEIGHT:	1156kg (2569lb)
MILEAGE:	10.0l/100km (28mpg)

Left: ***This is the diesel Ambassador, which is by far the most popular variant in its home market – that's despite it being one of the slowest cars ever made.***

The Ambassador's cabin is a confusing fusion of classic and modern. It's still just as cramped and narrow as the Morris Oxford, but has modern seat fabrics and dashboard dials. Well, modern-ish …

Morris mechanical components were dropped in favour of Indian-built power plants and, later, Japanese-supplied ones. The latest versions of the Amby have Mitsubishi engines, or a particularly unpleasant ex-Peugeot diesel.

While several aspects of the Ambassador have been updated for modern driving, one of the most essential items has been overlooked, and the brakes are as dreadful today as they were in 1954. In frenetic Indian traffic, it makes for interesting motoring …

The Amby is a classic shape. It started life as the Morris Oxford in 1954, and its shape continues unchanged more than half a century later, although a recent facelift has given the front end the appearance of a new Mini.

HUMBER SCEPTRE *(1967–76)*

The post-1967 Sceptre is a shining example of how American giant Chrysler completely misunderstood the Rootes Group when it took over the struggling British company in 1964. For years, Humbers had been upmarket cars, with plush trim levels, excellent build quality and a badge to which Middle England aspired almost as much as it did to Jaguar or Rover. The previous Sceptre, built between 1963 and 1967, offered all the traits Humber buyers loved: intricate styling, beautiful interior design and steadfast reliability. But Chrysler misinterpreted this, and went down the American route of badge-engineering, adorning a Hillman Hunter with an oversized grille, lashings of chrome trim, a dashboard coated in unconvincing fake wood and the ultimate ignominy of a vinyl roof. The Sceptre wasn't so much a bad car as a desperately cynical one, and it marked the end for one of Britain's finest marques.

SPECIFICATIONS

TOP SPEED:	163km/h (101mph)
0–96KM/H (0–60MPH):	13.6secs
ENGINE TYPE:	in-line four
DISPLACEMENT:	1725cc (105ci)
WEIGHT:	983kg (2185lb)
MILEAGE:	10.0l/100km (28mpg)

Left: ***It might have looked like a Hunter, but Chrysler tried to make the inside of the Sceptre look more upmarket. There was fake wood on the dashboard, chrome rings on the dials and fake leather seats – but posh it certainly wasn't.***

The Sceptre was better built than many cars in the 1970s, but when rust got hold it wouldn't go away. The front wings were first to go, tearing themselves away from the front panel in extreme circumstances.

Previous Sceptres had always been solid, pleasant cars to drive – so traditionalists swapping to the new one were in for a surprise. The new car's chassis was very twitchy, and, being lighter, the car was more prone to oversteer.

Most of the Sceptre was pretty bad, but the engine was its saving grace. The 1.7-litre (104ci) unit offered a decent amount of power for its size and was popular with performance fans, meaning many were sacrificed for custom projects.

KIA PRIDE *(1991–95)*

Yet another motoring misnomer – few people would have taken much pride in owning a tarted-up 1980s Mazda 121 with whitewall tyres and a roll-back sunshine roof. Yet Pride was the name Korean manufacturer Kia gave to the car it used to launch itself in American and European markets. The Pride was cheap and cheerful, with lots of shiny plastic inside and only one engine option – a coarse and unrefined 1.3-litre, four-cylinder unit that could trace its roots back to the late 1960s. Kia went for the no-nonsense marketing approach with the car, proclaiming it as an ordinary car for ordinary people. We doubt many ordinary people will have specified the optional metallic pink paintwork offered between 1992 and the model's eventual demise in 1995, yet, despite its many shortcomings, the Pride was a surprising sales success and established Kia in the global car market.

SPECIFICATIONS

TOP SPEED:	148km/h (92mph)
0–96KM/H (0–60MPH):	12.8secs
ENGINE TYPE:	in-line four
DISPLACEMENT:	1324cc (81ci)
WEIGHT:	803kg (1784lb)
MILEAGE:	6.7l/100km (42mpg)

Left: ***Kia showed only half of the car on the cover of the Pride brochure – was this a ploy to avoid putting people off?***

Park a Kia Pride next to a 1988 Mazda 121 and you'd be hard pushed to tell the difference. Some critics cruelly compared the optional roll-back roof to the top of a sardine can.

It looked the same as the Mazda 121 inside, but somehow Kia managed to make the cabin look even cheaper by using different grades of plastics and paler colours. It was a cheap car – and felt like it.

Only one engine was offered – a tough 1.3-litre (79ci) unit. It was fairly responsive, but the engine note was so loud you never felt like driving it especially quickly for fear of deafness.

The Korean car market was still young when the Pride appeared, and buyers had strange ideas of what counted as upmarket. It was the first car in two decades to come with whitewall tyres as standard.

LADA RIVA *(1969–present)*

Few cars have reached such levels of awfulness as the Lada Riva, but equally few have acquired such a cult following among European buyers. Built primarily as transport for the masses in the communist USSR, the Riva was also sold in Western Europe as a budget car, where it established a loyal repeat customer base, despite its atrocious ride, gutless overhead cam engine and handling characteristics that could best be described as terrifying. The car was a 1960s Fiat 124 under the skin, but was made out of cheap recycled metal with an engine and nasty four-speed gearbox designed and built to survive countless Russian winters. The Riva was a handful to drive, with heavy steering and a wayward rear end, but legions of pensioners loved its attractive pricing and mechanical simplicity. Today, there's a thriving owners club.

SPECIFICATIONS

TOP SPEED:	140km/h (87mph)
0–96KM/H (0–60MPH):	16.1secs
ENGINE TYPE:	in-line four
DISPLACEMENT:	1452cc (89ci)
WEIGHT:	982kg (2182lb)
MILEAGE:	10.0l/100km (28mpg)

Left: ***Cheap and durable, the Lada Riva was a surprisingly able rally car.***

By the 1990s, few cars were prone to rust. The Riva was the exception that proved the rule, and rust in the wings, sills and trunk lids was common. In fact, it had a propensity to rust like a tin can on a beach. Plastic trim also fell off regularly, often without provocation.

Amazingly, Lada designed and developed its own engine for the Riva. Even more remarkably, it was a relatively advanced overhead cam unit. It was not refined or fast or economical, but it was incredibly tough and rarely broke down.

European regulations dictated all cars built from 1993 onwards had to have a standard catalytic converter. Impressively, Lada did manage to comply, but if the convertor packed up the car was fit only for scrap because repairs were more than its value!

Throw a Lada into a bend, then wind on the steering lock and it cocks its inside front wheel in the air in dramatic fashion – but make sure there's nothing to hit, as it also surrenders all grip!

LONSDALE SATELLITE (1982–84)

New Zealand's only indigenous car manufacturer went on a mission to establish itself as a global player in the early 1980s, but few were fooled by the truly appalling Lonsdale. Based on the 1979 Mitsubishi Sigma and keenly priced, the car offered plenty of standard equipment and space for a large family to travel in comfort. But these were its only advantages.

Among its minus points were its lack of rust resistance and its unpleasant handling. By far the biggest problem, though, was the engine. The 2.6-litre powerplant was unique to the model and unusually large for a four-cylinder unit, delivering plenty of torque but little in the way of performance. It was also desperately unreliable – a huge problem because, when Lonsdales expired, as they inevitably did, overseas owners had to wait months for the replacement parts to arrive from New Zealand.

SPECIFICATIONS	
TOP SPEED:	173km/h (107mph)
0–96KM/H (0–60MPH):	10.8secs
ENGINE TYPE:	in-line four
DISPLACEMENT:	2555cc (156ci)
WEIGHT:	1165kg (2588lb)
MILEAGE:	11.8l/100km (24mpg)

Left: *Two body styles of the Lonsdale were offered, of which the station wagon was the most popular. That said, popular isn't really the right word, as a glance at the sales figures will confirm.*

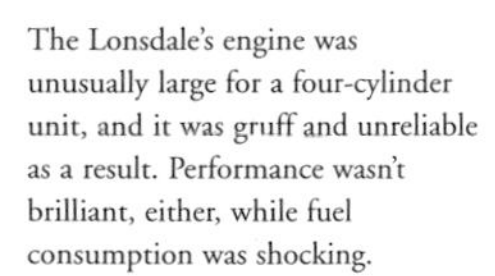

The Lonsdale's engine was unusually large for a four-cylinder unit, and it was gruff and unreliable as a result. Performance wasn't brilliant, either, while fuel consumption was shocking.

At least the Lonsdale was practical. In station wagon form, it boasted a huge luggage area, while accommodation for passengers was surprisingly comfortable. But being practical doesn't make it a good car.

Like many cars of its era, the Lonsdale was prone to corrosion, with rot manifesting itself in the front and rear wheel arches, sills and inner wings rather more quickly than was acceptable.

Styling wasn't the Lonsdale's strong point: it was based on the 1980s Mitsubishi Sigma, and, apart from the badges, the only difference between the two was the design of the rear tail-light clusters.

MAHINDRA INDIAN CHIEF *(1993–95)*

The Mahindra was sold as a utility vehicle in its home market of India. However, the company saw the growth of 4x4s in Europe and America, and decided to try to crack these markets with this model in the early 1990s. The move was ultimately a failure, but the Indian Chief could easily have been a success if it were better planned. In concept, it was a great idea. The car was the original Jeep CJ – an American legend – and it was built under licence, primarily for the Indian armed forces. It looked the part, especially in black with chrome alloy wheels, but its appeal was lost as soon as you settled into the driver's seat.

The cabin tended to fall apart within days of delivery being taken, while the 2.1-litre Peugeot diesel engine was far more suited to the van it came from, offering a top speed little over 112km/h (70mph). Handling was awful, and Mahindra's UK importer once admitted it had to practically rebuild every car before selling it to a customer, such was the dire build quality.

SPECIFICATIONS

TOP SPEED:	105km/h (65mph)
0–96KM/H (0–60MPH):	33.9secs
ENGINE TYPE:	in-line four diesel
DISPLACEMENT:	2112cc (129ci)
WEIGHT:	1328kg (2951lb)
MILEAGE:	9.4l/100km (30mpg)

Left: ***The Mahindra name is synonymous with extremely cheap, but even more awful utility vehicles.***

The leaf springs and live rear axle made handling unpredictable and gave the Mahindra shocking ride quality, and little in the way of shock absorption meant that every road bump was transferred from the wheels to the driver's spine.

The Indian Chief was supposed to be tough, and, although it would last a long time mechanically, that toughness didn't extend to the fixtures and fittings. Light covers and door hinges fell out, while the interior fell apart quickly.

Power, if you could call it that, came from a Peugeot-sourced commercial vehicle engine. The 2.1-litre (128ci) diesel unit had a reputation for longevity, but it was noisy, smelly and dreadfully slow.

Contemporary road testers expressed serious concerns about the Mahindra's steering, which lacked any kind of precision, was unduly heavy and required far too many turns from lock-to-lock.

MG MAESTRO *(1983–92)*

In 1981, the final MGB rolled off the production line, and fans of the illustrious marque thought the MG name was dead and buried. But if they were crying into their real ale when the MGB died, they must have felt positively suicidal when Austin-Rover revived the name for its hot hatch version of the Maestro in 1983. The MG Maestro was meant to be the British firm's answer to the VW Golf GTI – but it was upright and ugly, and the engine was awful, lacking refinement and, essentially, any kind of performance.

The MG was distinguished from lesser Maestros by its red seatbelts and 'cheese grater' alloy wheels; however, traits the MG Maestro had in common with its Austin sibling were an irritating voice synthesizer and rampant corrosion, especially around the rear wheel arches.

The original 1.6 version was replaced by the much-improved 2.0Efi in 1985, but even that was horrid. The final MG Maestro Turbo of 1989 was genuinely quick and quite good fun to drive, except when you tried to brake …

Left: ***MG purists sobbed into their real ale when Austin-Rover slapped their beloved badge onto the hideously ugly Maestro.***

SPECIFICATIONS	
TOP SPEED:	179km/h (111mph)
0–96KM/H (0–60MPH):	9.6secs
ENGINE TYPE:	in-line four
DISPLACEMENT:	1598cc (97ci)
WEIGHT:	968kg (2150lb)
MILEAGE:	10.0l/100km (28mpg)

MG versions of the Maestro differed from standard cars in a number of ways. Sporty touches included red seatbelts and what Austin-Rover called a 'techno-finish' to the instrument binnacle, and it also got a sporty three-spoke steering wheel.

Warnings on the voice synthesizer included 'low oil pressure', 'fuel level low', 'please fasten your seatbelts', 'check coolant level' and, rather bizarrely, 'the car is on its roof'.

Austin-Rover's 1.6-litre (98ci) R-Series engine was never the right choice for a performance car. It wasn't eager, and sounded terrible at high revs. Acceleration was a joke compared to rivals such as the VW Golf GTI.

Like the standard Maestro, the MG was a complete rot box. The rear wheel arches were first to go, followed by the tailgate, sills and front valance.

MORRIS ITAL (1981–84)

In 1998, Italdesign issued a book to celebrate its thirtieth anniversary. The only model missing from the publication was the one car to bear the name of the famed Italian styling house – the Morris Ital. And it's easy to see why. While models such as the Fiat Panda and Volkswagen Golf are rightly regarded as design classics, the Guigaro-restyled Morris most certainly isn't. It's a reskinned Morris Marina and, if anything, it's worse than the car it was supposed to replace. When you consider that the Marina itself was pretty awful, that's saying something. Under the skin, the Ital was identical to the Marina, which in turn was based on the 1948 Morris Minor. No prizes, then, for guessing that it was horrid to drive. Italdesign's facelift took away the Marina's stylish rear lights and replaced them with huge plastic units, while the front looked like a Chrysler Avenger – and, as we've already explained, that's nothing to boast about.

Left: ***The Ital station wagon was also pretty hideous, but at least it was moderately useful. Until it broke down.***

SPECIFICATIONS

TOP SPEED:	147km/h (91mph)
0–96KM/H (0–60MPH):	15.2secs
ENGINE TYPE:	in-line four
DISPLACEMENT:	1275cc (78ci)
WEIGHT:	2070lb (932kg)
MILEAGE:	33mpg (8.6l/100km)

The Ital's dashboard was revised so that the radio and heater controls were mounted on the centre console. For some strange reason, this pointed towards the passenger seat, making the controls invisible to the driver. Not clever.

You could have three different engines in your Ital. The 1.3-litre (79ci) used the ancient but characterful A-Series unit, which was at least fairly reliable. The 1.7-litre (104ci) and 2.0-litre (122ci) models got the atrocious O-Series that was introduced in the Princess and which was prone to oil seal failure.

Of all the Marinas offered, the Coupé was by far the most stylish. So when Morris developed the Ital, it decided, of course, to drop the attractive car, and concentrate on the dishwater-dull station wagon and saloon models instead.

In profile, it was difficult to tell the Ital apart from the Marina, although it did get wraparound indicators at the front and its own exclusive wheel trims. As redesigns go, though, it wasn't exactly adventurous.

PANTHER RIO *(1976–78)*

In theory, a truly upmarket version of the much-loved Triumph Dolomite would have been a great idea – but Panther's interpretation of the concept was ludicrous. The Rio was effectively a Dolomite Sprint, but with hand-forged alloy bodywork, Ford Granada headlights and a Rolls-Royce style front grille. Inside, the original seats were replaced by studded leather ones, and the wood on the dashboard was polished. In other words, it was a tasteless piece of 1970s kitsch.

As if all of that were not bad enough, Panther decided to charge three times the price of a standard Dolomite Sprint for its top-of-the-range Especiale version, making it only marginally cheaper than a genuine Rolls-Royce. To say the Rio was a flop would be an understatement – only 38 were made in a two-year production run.

Left: ***The sepia-toned cover of the Panther Rio brochure made it look like the sleeve of a 1970s record album!***

SPECIFICATIONS

TOP SPEED:	185k/mh (115mph)
0–96KM/H (0–60MPH):	9.9secs
ENGINE TYPE:	in-line four
DISPLACEMENT:	1998cc (122ci)
WEIGHT:	no figures available
MILEAGE:	10.0l/100km (28mpg)

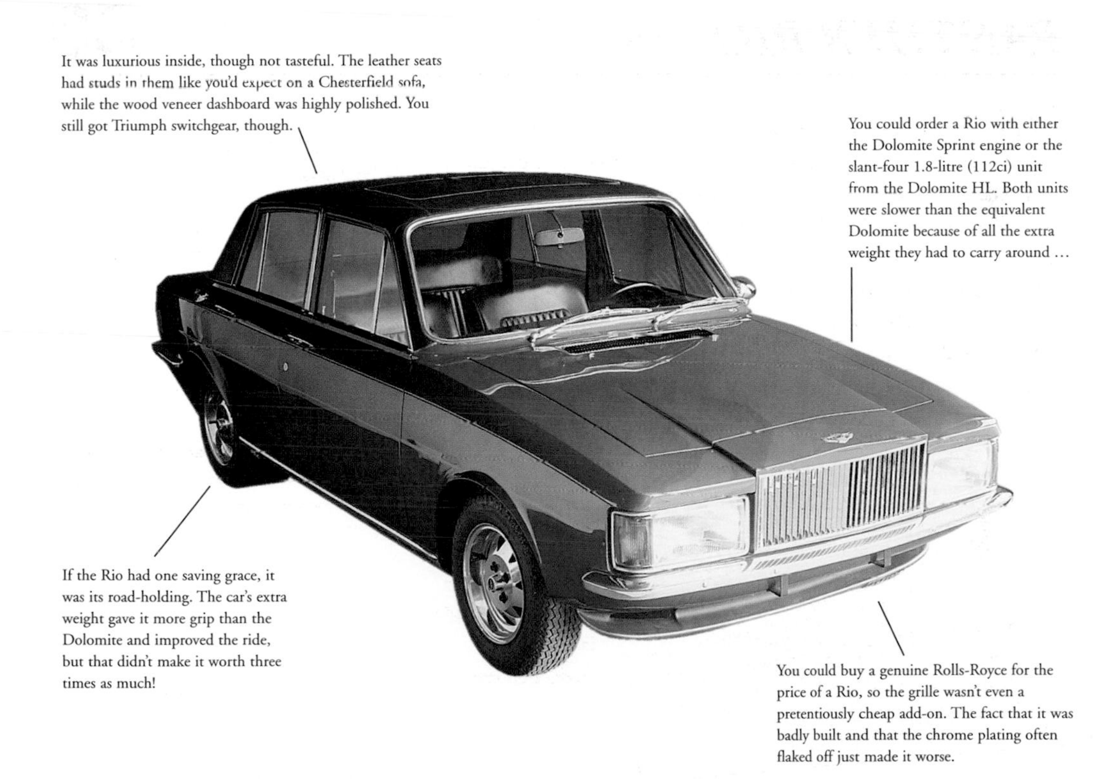
It was luxurious inside, though not tasteful. The leather seats had studs in them like you'd expect on a Chesterfield sofa, while the wood veneer dashboard was highly polished. You still got Triumph switchgear, though.
You could order a Rio with either the Dolomite Sprint engine or the slant-four 1.8-litre (112ci) unit from the Dolomite HL. Both units were slower than the equivalent Dolomite because of all the extra weight they had to carry around …
If the Rio had one saving grace, it was its road-holding. The car's extra weight gave it more grip than the Dolomite and improved the ride, but that didn't make it worth three times as much!
You could buy a genuine Rolls-Royce for the price of a Rio, so the grille wasn't even a pretentiously cheap add-on. The fact that it was badly built and that the chrome plating often flaked off just made it worse.

PAYKAN *(1976–present)*

When it was new, the Hillman Hunter was a respectable family car. Yes, it was a little rust-prone, but its simple mechanics and proven reliability gave it quality better than most of its rivals. That made it very attractive in later years for newly developing car markets and, when Chrysler pulled the plug on the Hunter in 1976, it was quick to sell the body pressings off to the Iranian government, which in turn used them to build the Paykan (pronounced 'Pie-can').

With austere specification levels and heavy-duty springs, the car was never a particularly pleasant thing to drive, and the majority of owners resorted to creative methods of keeping them on the road, meaning the remaining fleet of Paykans in Iran – and there are quite a few of them – are held together with bits of wood, string and carpet tape.

SPECIFICATIONS

TOP SPEED:	139km/h (83mph)
0-96KM/H (0-60MPH):	17.8secs
ENGINE TYPE:	in-line four
DISPLACEMENT:	1496cc (91ci)
WEIGHT:	915kg (2035lb)
MILEAGE:	915kg (27mpg)

Left: *For years, Paykan didn't even have to advertise. Now the Iranian car market is more open to imports, glossy brochures are the norm. Despite tough competition, the Paykan is still a good seller.*

There's no hiding the Paykan's origins. Despite having its own design of front and rear lights, the silhouette of the Hillman Hunter is instantly recognizable behind the car's unique façade.
Only one specification level was offered – and it was basic. Buyers got vinyl seats, a plastic dashboard and gauges to show fuel level, temperature and speed – but that was it.
Although based on the Hillman Hunter, the Paykan had its own rear springs, engineered to cope with rough Iranian roads. They were much firmer than the Hunter's, giving a horrible, bouncy ride.
The Paykan wasn't immune from the rust troubles that plagued the Hillman Hunter in Britain, although the dry Iranian climate meant that the rot took longer to manifest itself. Even so, inner and outer wings and rear wheel arches were vulnerable.

PERODUA NIPPA *(1996–2000)*

Some cars really don't deserve to live on, and the Daihatsu Mira was one of them. A simple but largely unpleasant city car, it served a purpose in Japan's congested streets, where its tiny wheelbase, low running costs and 847cc engine made it reasonably popular. But when the car came to the end of its life, its design was sold by Daihatsu to Malaysian company Perodua, which stripped it of any creature comforts and sold it as a budget model.

The 1980s-throwback Nippa was a truly awful machine, which came with no radio, no heated rear window and a terrifying habit of veering across the road in crosswinds. Perodua exported the car to both the UK and South Africa, where it was the cheapest model available on the market. Anyone who drove one soon found out why that was so …

SPECIFICATIONS

TOP SPEED:	129km/h (80mph)
0–96KM/H (0–60MPH):	21.1secs
ENGINE TYPE:	in-line three
DISPLACEMENT:	847cc (52ci)
WEIGHT:	678kg (1506lb)
MILEAGE:	5.7l/100km (50mpg)

Left: ***According to Perodua, 'size is everything'. That's a very strange thing for a company to say when it's trying to market a tiny, ancient, badly built ex-Japanese city car.***

Don't expect any luxuries inside a Nippa. You get a speedometer and a fuel gauge, seats, a gear lever and an ashtray. 'Luxury' items, such as a radio and a lid for the glove box, were optional extras.

You don't get much in the way of power from the Nippa – it will accelerate to a reasonable cruising speed, but only if you persevere. And once you get there, it's noisy and unrefined because a lack of soundproofing was one way costs were kept to a minimum.

The Nippa's light weight and upright shape meant it struggled to cope in blustery weather. A strong crosswind was enough to make the car drift across a road, often with positively terrifying consequences.

The short wheelbase and tiny wheels gave the Nippa a disturbingly bumpy ride, with a tendency to bounce around on uneven surfaces and transfer jolts from bumpy surfaces directly through to the car's cabin.

PLYMOUTH CRICKET (1973–80)

When General Motors took on the fuel crises by making an American version of the Vauxhall Chevette, the Chrysler Corporation decided to do something very similar. The model in question was the Plymouth Cricket – a federal-spec version of the trusty but uninspiring Hillman Avenger. But few American buyers were impressed by either the model's cramped interior or its ability to rust rather too quickly.

Power came from the Avenger's proven 1.6-litre pushrod engine, which served British buyers well. But with its carburettor detuned to satisfy US emissions legislation, it never ran properly and was mind-numbingly slow. The styling alterations, including huge impact fenders and amber marker lights, didn't do the Avenger's lines any favours either. Today, the Cricket is largely forgotten.

SPECIFICATIONS

TOP SPEED:	134k/mh (84mph)
0–96KM/H (0–60MPH):	19.8secs
ENGINE TYPE:	in-line four
DISPLACEMENT:	1295cc (79ci)
WEIGHT:	850kg (1889lb)
MILEAGE:	8.8l/100km (32mpg)

Left: ***Using the psychedelic imagery of the era, the adertisers attempted to portray the Plymouth Cricket as the colourful and wacky option for the discerning driver. However, the car was a failure in the competitive North American automobile market.***

Safety laws meant changes to the interior, with chunks of PVC-coated foam attached to the dashboard and racy high-back seats – but the modifications looked like afterthoughts, which in essence they were …

Rust was a problem, too. It was common around the front wings, valance panel and rear spring hangers, while the floorpans were also susceptible.

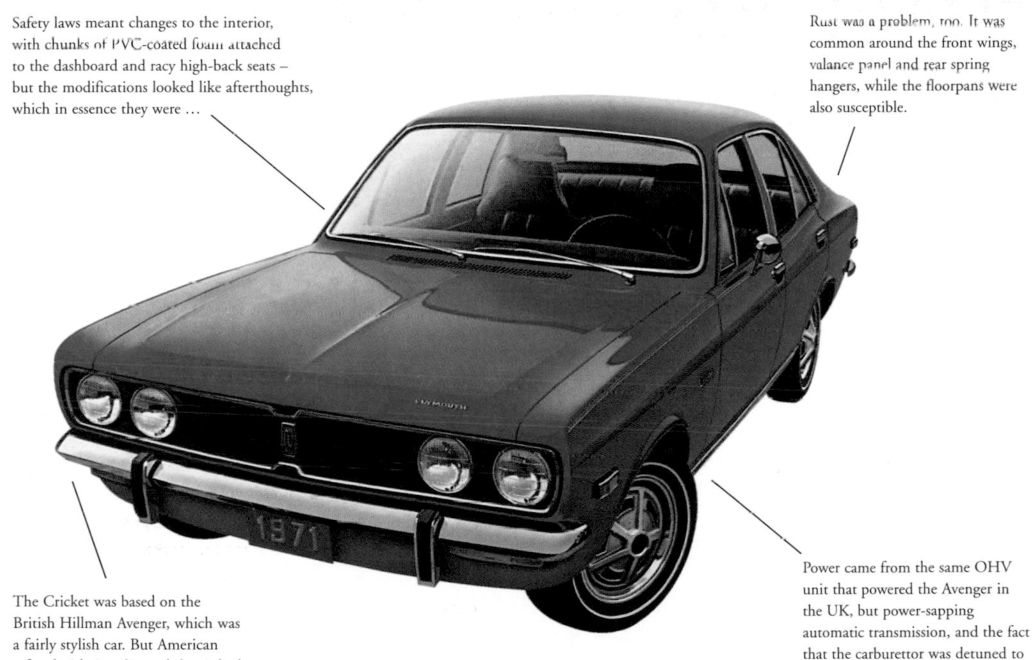

The Cricket was based on the British Hillman Avenger, which was a fairly stylish car. But American safety legislation dictated that it had to have thicker fenders, which ruined the car's lines.

Power came from the same OHV unit that powered the Avenger in the UK, but power-sapping automatic transmission, and the fact that the carburettor was detuned to comply with emissions regulations, made the Cricket horribly slow.

PORSCHE CAYENNE (2003–present)

From a driver's perspective, it's true, there's nothing wrong with the Porsche Cayenne. It's powerful, has remarkable handling for an SUV and is well built. But there's so much wrong with the ideology behind it that Porsche purists despise the model – and not without good reason. The most obvious flaw is the styling. It's far from elegant, with an awkward mix of traditional Porsche styling cues and the slab-sided profile that's endemic of any off-road machine.

Then there's the cynicism – the Cayenne was built so that people could have lots of kids and still own a Porsche, which is hardly the reason so many owners scrimp and save to get their hands on one. Finally, it was developed alongside the VW Touareg and even shares its entry-level six-cylinder engine. Unless you have to have the Porsche badge, the VW is a far better buy all round.

SPECIFICATIONS

TOP SPEED:	266k/mh (155mph)
0–96KM/H (0–60MPH):	5.6secs
ENGINE TYPE:	V8 turbo
DISPLACEMENT:	4511cc (275ci)
WEIGHT:	2355kg (3650lb)
MILEAGE:	13.4l/100km (21mpg)

Left: *The cutaway drawing clearly shows the Cayenne's springing system and transmission layout, both of which are identical to those offered on the VW Touareg.*

The Cayenne claims to be a Porsche – but the vehicle has the same doors as the VW Touareg, and under the skin is essentially the same architecture.

V8-engined Cayennes have a power plant that's purely Porsche designed – and it's a cracker. But entry-level cars come with a 3.2-litre (195ci)) Volkswagen V6 and, despite the Porsche badging, they're not even that fast …

Whatever angle you look at it from, the Cayenne's styling is challenging. The lines aren't harmonious, and some writers have compared the car's bulbous nose and huge air intake to the face of a wide-mouthed toad.

For a car so large and tall, the Cayenne has good levels of grip and assured handling. But the downside is an incredibly harsh ride, which transmits every jolt from the suspension upwards into the cabin.

PROTON WIRA *(1995–2004)*

Yet another rehashed old Japanese machine: the Proton Wira debuted in 1995 and was based on the Mitsubishi Lancer of four years previous. Built in Malaysia, the Wira was exported to the UK, South Africa and Australia, where it achieved moderate success as an attractively priced and spacious saloon. Reliability was good, and most are still going strong. So why is it in this book? Quite simply, it's one of the most soulless, bland and dynamically tedious cars ever made.

The Wira has no styling panache, a complete lack of handling finesse, an interior packed with cheap, horrible plastics, a bumpy ride, a rubbery gearshift and an engine that booms and roars but provides mediocre performance. To anyone with even a passing interest in cars, the Wira is a complete travesty – it turns the automobile into a consumer durable, and has about as much character as a washing machine.

SPECIFICATIONS

TOP SPEED:	182km/h (113mph)
0–96KM/H (0–60MPH):	10.4secs
ENGINE TYPE:	in-line four
DISPLACEMENT:	1597cc (97ci)
WEIGHT:	1063kg (2362lb)
MILEAGE:	35mpg (8.1l/100km)

Left: ***Proton made a big fuss about the special edition Wira having 8 per cent more horsepower. But in reality it was a negligible gain of around 7bhp more than the standard car, and was hardly noticeable from behind the wheel – until you stared at the garish red dials on the dash.***

Don't expect the interior to have any kind of feel-good factor. The plastics used are cheap and brittle, while the PVC-coated rear sun visors are especially unpleasant. Unless you have a fetish for light grey, it's also unbearably dull.

Power, if you can call it that, comes from a choice of 1980s Mitsubishi-sourced engines of either 1.3-litre (79ci) or 1.5-litre (92ci) capacity. The units are old-fashioned, gruff and unresponsive, with plenty of noise at high revs.

Handling is very ordinary. The Wira lacks grip and is prone to understeer, while the steering lacks any kind of feel and body roll in corners is dramatic. The ride quality is also dreadful, especially for rear-seat passengers.

There are hardly any redeeming features in the Wira's body design – it lacks personality, with bland lines and a tacky finish. Hatchback models also have ugly rear-end styling, with their own unique light units.

ROVER 800 (1986–99)

Motoring history is packed with missed opportunities, but the Rover 800 has to be one of the biggest. Launched to an expectant public in 1986, Britain's new executive car was supposed to take on the likes of BMW and Mercedes. A tie-in with Honda meant the Japanese firm would provide the V6 engines for the 800 and its sister car, the Honda Legend, while Rover itself would build the four-cylinder units. What happened next was painfully predictable. The Rover engines packed up, along with the Lucas electrical systems, while the V6 engines were, in true-Honda style, high-revving and sporty, making them rather inappropriate for a luxury model. Not only that, but build quality was terrible as well, with horrific rust and dashboards that curled up at the edges in hot weather. Although the 800 of 1991 got a facelift and was an infinitely better car, its reputation was for ever sullied.

Left: ***You could even buy a range of tasteless body kits for the Rover 800 range, which stopped the car looking bland and made it look daft instead.***

SPECIFICATIONS	
TOP SPEED:	187km/h (116mph)
0–96KM/H (0–60MPH):	10.2secs
ENGINE TYPE:	in-line four
DISPLACEMENT:	1994cc (122ci)
WEIGHT:	1260kg (2800lb)
MILEAGE:	10.0l/100km (28mpg)

Early Rover 800s didn't even look good, with styling that resembled an engorged Austin Montego. Post-1991 cars were restyled and looked much better, but they were still rather boring compared to executive rivals.

It was supposed to be an executive car, but the 800 seemed to have been put together out of the Austin-Rover parts bin. That meant it had a myriad of gauges and switches plucked from more downmarket cars in the company's line-up, and looked weak against German rivals.

Rust affected early 800s quite badly, with many cars showing signs of corrosion at less than five years old – leading to many warranty claims for the manufacturer. The doors, trunk floors and rear wheel arches were always first to go.

Rover offered its own twin-cam 2.0-litre (122ci) engine in the 800, which cost millions of pounds to develop. But it was difficult to maintain and repair, which was unfortunate given its dreadful reliability record.

ROVER CITYROVER (2003–present)

With buyers in Europe turning more and more towards smaller models, British manufacturer Rover decided to re-enter this market segment in 2003. But rather than develop a new model, which was way beyond the struggling firm's budget, it signed a manufacturing agreement with Indian engineering firm Telco. In its home market, Telco sells the Tata Inidica – the most successful Indian model ever. It is a reasonably modern small hatchback, with pleasant styling and a fairly spacious interior.

For Rover, it seemed the perfect tonic, so the car was introduced in Europe as the CityRover. Sadly, though, it was not modified enough to suit European tastes. Its low-rent plastics and cheap-feeling trim did it little favours, and the noisy engine and awful gearshift didn't help either. Rover expected buyers to pay the same as they would for an array of superior rivals, but few customers were fooled into doing this.

SPECIFICATIONS

TOP SPEED:	168km/h (106mph)
0–96KM/H (0–60MPH):	11.8secs
ENGINE TYPE:	in-line four
DISPLACEMENT:	1396cc (85ci)
WEIGHT:	1020kg (2266lb)
MILEAGE:	6.9l/100km (41mpg)

Left: ***The young couple smooching by the wall are waiting until nobody else is looking. It's their CityRover in the foreground, and they don't want to be seen getting into it ...***

Smart outside, but horrible inside – that's the CityRover. The dashboard feels cheap and the instruments sit in a hastily attached box on top of the dash. The interior light pops out in your hands and the column stalks feel as if they might snap if you flick them too hard.

The most successful aspect of the CityRover is its styling. It looks fairly smart, with lines that better quite a few rivals. But behind the chic veneer it's a cheap and rudimentary car, and Rover's pricing made it poor value for money.

The CityRover's engine is based on an old Peugeot unit and is fairly responsive, but it's not very environmentally friendly and it's attached to a slack, inaccurate gearbox that's deeply unpleasant to use.

Handling is actually quite good – but that doesn't make the CityRover a good car. It was also criticized at launch for not having anti-lock brakes as standard, an oversight in the modern marketplace.

SEAT MARBELLA *(1988–97)*

If you've ever taken a holiday in Spain, the Balearic Islands or the Canaries, the chances are that you've driven a SEAT Marbella. For years, it was the budget hire car of choice in Southern Europe, with its basic interior and flimsy, tinny bodywork. The car was based on the 1980 Fiat Panda and was launched in 1988, as the Spanish firm's first major contender in the booming supermini market. But in reality, it was outdated as soon as it came out, and the awful build quality, queasy colour schemes and cartlike ride did it no favours whatsoever. In its defence, the Marbella was a cheap car with low running costs – but once they'd sampled it, most buyers decided it was worth paying a little extra to avoid the experience of driving one on a day-to-day basis.

SPECIFICATIONS

TOP SPEED:	131km/h (81mph)
0–96KM/H (0–60MPH):	18.7secs
ENGINE TYPE:	in-line four
DISPLACEMENT:	903cc (55ci)
WEIGHT:	706kg (1569lb)
MILEAGE:	6.3l/km (45mpg)

Left: ***The Marbella was awful, so quite why this driver chose to risk injury and take one rallying is beyond our comprehension ...***

There were no creature comforts inside the Marbella. The fabric seats were lightly padded, while the dashboard was also made out of a simple cloth material. The doors had a painted metal finish, and it had rubber mats instead of carpets.

SEAT gave it a new nose, but there was no hiding the fact that the Marbella was an early Fiat Panda under the skin. In profile, there was little to tell the two cars apart, with the same boxy styling and flat windshield as the Italian car.

Fiat updated the Panda in 1986, but SEAT didn't follow suit. As a result, the Marbella still had cart springs at the rear, which in turn gave it an incredibly bouncy ride and somewhat unpredictable handling.

While not as fragile as the original Fiat Panda, the Marbella was still prone to rust if not properly maintained. The paint finish was poor and often faded, while rust found its way into the rear wheel arches, doors and trunk floor.

TRIUMPH ACCLAIM (1981–84)

The Acclaim was a critical car for the British motor industry, as it marked the first tie-in between what was then British Leyland (later Rover) and Japanese maker Honda – an alliance that would go on to save the company in the following decade. But the Acclaim certainly wasn't the car to stop the rot that was the result of years of pandering to trade unions. It was based on the Honda Ballade, a four-door saloon popular in Japan and America. If BL had badged the car as an Austin or Morris, it would probably have been a success, but instead they chose to put the mark of upmarket brand Triumph on the nose and tail.

Most Triumph buyers were used to luxurious small cars, and the Acclaim's plastic cabin and bland Japanese styling just weren't their cup of tea. The model was scrapped after two years.

SPECIFICATIONS

TOP SPEED:	148km/h (92mph)
0–96KM/H (0–60MPH):	12.9secs
ENGINE TYPE:	in-line four
DISPLACEMENT:	1335cc (81ci)
WEIGHT:	784lb (803kg)
MILEAGE:	34mpg (8.3l/100km)

Left: ***'The Acclaim beats all comers,' says Triumph, referring to such mighty adversaries as the Renault 9 and Ford Escort Mk 3, but conveniently forgetting good cars such as the VW Golf and anything Japanese.***

Triumph purists liked rear wheel drive and firm suspension, neither of which the Acclaim could offer. It had soft springs, front-wheel drive and little in the way of handling prowess, with a chassis prone to understeer and sloppy steering response.
Triumphs were always upmarket and well appointed, so the Acclaim was a disappointment to the brand's traditional buyers. Gone were the upmarket wood and leather, to be replaced by velour upholstery and a plastic dashboard.
Although it had Japanese reliability engineered in, the Acclaim could still rust in the same way that British cars always had done. So the rear wheel arches rotted out, the doors went crusty and the sills gave way.
Power was provided by tough and reliable Honda engines, so the news wasn't all bad. But Triumph offered its own Trio-Matic three-speed auto transmission, which was especially slow to respond and delivered dire performance.
AEL 442Y

VANDEN PLAS 1500 (1974–81)

As if the Austin Allegro were not a cruel enough joke to play on the car-buying public, the Vanden Plas 1500 was pure evil. With a name steeped in coach-building tradition, Vanden Plas had been making upmarket versions of British Leyland cars for years. In 1974, BL's top brass decided that the Allegro should be its next project – the aim being to ensnare retirees used to the accoutrements of a luxury car, but wanting something smaller and cheaper to run. Herds of Allegros were pulled off the production line and trailered to Vanden Plas's factory in Kingsbury, North London, where they lost what was left of their dignity to a burr walnut dashboard, picnic tables, leather seat facings, Wilton carpets and a bolt-on grille. Truly pretentious, and no more reliable than the troublesome Allegro on which it was based.

SPECIFICATIONS

TOP SPEED:	144k/mh (90mph)
0–96KM/H (0–60MPH):	14.5secs
ENGINE TYPE:	in-line four
DISPLACEMENT:	1485cc (91ci)
WEIGHT:	900kg (2000lb)
MILEAGE:	9.4l/100km (30mpg)

Left: ***From the back, you'd have thought this was just an ordinary Austin Allegro. Little did you know what spectacular treats awaited you inside. Picnic, anyone?***

The Vanden Plas 1500 was an Austin Allegro in all but name. At the front, it featured a very silly radiator grille that looked far too large for the car to which it was attached, yet when Rover bolted similarly incongruous protuberances to the front of Hondas 20 years later, nobody thought it was funny …

Once you're inside a VP 1500, you can forget how daft it looks outside. The leather seat facings, picnic tables, walnut dashboard and Wilton carpets are truly luxurious. So it is, at least, pleasant to be a passenger, if not a driver!

Just like the Allegro, the VP 1500's luggage hold wasn't the best place to store things. Thanks to fundamental design flaws, the seals always leaked, leading to a large puddle inside and resultant corrosion in the trunk floor.

The Vanden Plas 1500 was offered with British Leyland's E-Series engine. The overhead cam unit was fairly quick, but unrefined and prone to cam chain failure and oil starvation. With the auto gearbox option, it was also dreadfully slow.

WOLSELEY SIX 18-22 SERIES *(1975–76)*

The Vanden Plas 1500 wasn't the only horror foisted on Middle England by British Leyland's cost-cutting management. The Princess 'Wedge' debuted in 1975, and this was the ultimate version, wearing the badge of the fabled upmarket car builder Wolseley.

Buyers of the marque were used to supreme quality and lavish luxury, and it's true that the 18-22 Series offered plenty of standard equipment, including a wooden dashboard and a vinyl roof, but few were prepared to accept the Wedge's avant-garde 'styling', nor did they like its tendency to munch its way through engine mounts and drive shafts, while slowly listing to one side as the gas-filled suspension depressurized itself. For Wolseley fans, to see their revered marque's illuminated grille badge shine from the Wedge's nose was sacrilege enough, but then came British Leyland's decision to drop the Six, and the Wolseley name entirely, after less than a year.

SPECIFICATIONS

TOP SPEED:	166k/mh (104mph)
0–96KM/H (0–60MPH):	13.5secs
ENGINE TYPE:	in-line six
DISPLACEMENT:	2227cc (136ci)
WEIGHT:	1187kg (2638lb)
MILEAGE:	11.3l/100km (25mpg)

Left: ***The rear was by far the Wolseley Wedge's best aspect. Wolseley models also got unique badging.***

As with the Princess, the Wolseley 'Wedge' was fairly rust-prone, with rot working its way into the rear wheel arches, sills and door bottoms at alarming speed. Owners weren't impressed.
Harris Mann's famous 'Wedge' was an avant-garde design that was loved or loathed. Usually loathed by fans of the traditionally British Wolseley marque.
Spot a Wolseley by its thick-pile carpets and genuine wood veneer dashboard, as opposed to the fake stuff offered in Austins and Morrises. It also had a traditional Wolseley trademark – a grille badge that lit up with the car's headlights.
The Wolseley only came with the 2.2-litre (134ci), six-cylinder engine offered in the most upmarket Austin and Morris variants of the Princess. It was fairly quick, but liked to burn oil, and the engine mounts and drive shafts gave way regularly.
POC 804N

CAR COMFORT FOR 5

at a total running cost of less than 1d. per mile

THE FAMILY SAFETY MODEL

Designed and Built for 2 adults and 3 children with full weather protection for all

Manufactured by

SHARP'S COMMERCIALS LTD., (Est. 1922) PRESTON, LANCASHIRE

MOTORING MISFITS

Some cars are truly awful, yet it is impossible to pinpoint why. They are either out of place because they are so obscure, or – worse – are completely wrong in almost every respect. Cars such as the Amphicar, Bond three-wheeler, Suzuki X90 and Marcos Mantis are completely off the wall. They are motoring monstrosities: cars that should never have seen the light of day, but did so because somewhere along the line at least one person believed their place in the market was justified. Many of them are historically interesting and have even acquired something of a cult reputation.

Other cars are included because it would be impossible to put them in any of the other categories in this book. They are awful in so many ways that they could have fallen into any of the previous sections, but their various faults are impossible to separate. Cars such as the AMC Eagle, Nash Metropolitan and Wartburg Knight, which were badly built, poorly designed and financially disastrous for their makers, nestling among legions of other motoring misfits.

Left: *It might have been able to do a U-turn in its own axis, but the Bond Minicar had absolutely no other redeeming features.*

ALFA ROMEO 6 (1980–85)

It might have had a fine reputation for building attractive sports cars, but Alfa Romeo didn't really get a grip on how to make large saloon models until it introduced the stunning 164 in 1988. Its predecessor was somewhat less illustrious. Based on an extended Alfetta platform, the Alfa 6 was not only hideous, but also plagued by electrical demons, rampant body rot and serious reliability issues. And if its ugliness already made it an eyesore in the Alfa line-up, then the fact that it suffered from appalling handling and a dreadful gearbox and was not even especially quick made it even more of a misfit among the Italian firm's line-up of otherwise great drivers' cars. A dismal attempt at an executive car, and one that deserves to be as widely forgotten as it has become.

SPECIFICATIONS

TOP SPEED:	185km/h (115mph)
0–96KM/H (0–60MPH):	11.4secs
ENGINE TYPE:	V6
DISPLACEMENT:	2492cc (152ci)
WEIGHT:	1539kg (3241lb)
MILEAGE:	12.3l/100km (23mpg)

Left: ***Alfa claimed the 6's body was strong and practically rust-resistant in the brochure, but both claims were proved incorrect.***

Alfas of this era were prone to excessive corrosion, as was the 6: structural problems were common in the sills, rear suspension mounts and front inner wings.

Alfa Romeo has always been known for its design flair – but the Alfa 6 was a very notable exception. It lacked any kind of design innovation, and the front-end treatment was especially bland.

The Alfa 6's gearbox was mounted on the rear axle, giving an obstructive gear change. It also caused weight distribution problems, as the car was heavier at the back than the front.

AMC EAGLE WAGON (1980–84)

In many senses, the AMC Eagle Wagon was years ahead of its time. Look at most manufacturers' model lines today, and you'll find a version of one of its standard cars on pumped-up suspension with chunky plastic body armour. The Eagle, then, was a good idea – unfortunately it was one that buyers simply couldn't get their heads around. Maybe it would have been a success if it had been a half-decent car. But it wasn't, and that was definitely the Eagle's biggest shortcoming.

For starters, AMC took the already terrible Gremlin as the basis, and gave it extra bodywork and chunky protective mouldings that made it look even more repulsive. It also had an old, inefficient six-cylinder motor and a clunky, three-speed automatic gearbox licensed from Chrysler. In combination, they made the Eagle a spectacular failure.

SPECIFICATIONS

TOP SPEED:	145km/h (90mph)
0–96KM/H (0–60MPH):	16.8secs
ENGINE TYPE:	in-line six
DISPLACEMENT:	4229cc (258ci)
WEIGHT:	1492kg (3283lb)
MILEAGE:	14.1l/100km (20mpg)

Left: *The Eagle's cabin was pure 1980s Americana. Acres of tan coloured vinyl and cheap plastic were sadly typical of the era.*

The Eagle Wagon was supposed to be an innovation for the 1980s, but AMC stripped it of any technical advantage by using an old, slow and inefficient engine, coupled to a crude transmission.
Practicality was the Eagle's greatest plus point, as it made for well equipped and spacious family transport. But given the model's many foibles, this was not a good reason for buying one.
It was the Eagle Wagon's styling that put many buyers off. The car was based on the Gremlin, and the jacked-up suspension and fake wood side panelling did it no favours.

AMPHICAR (1961–68)

Germany has one very small coastline, right up in the freezing cold north of the country. So quite why its creator, a slightly eccentric engineer called Hans Trippel, thought his Amphicar would entice German buyers is anyone's guess. But Trippel's barmy creation was nothing if not ingenious, and was designed to be just as capable in water as it was on the road. In reality, it wasn't a success. For starters, any car that's supposed to be capable of travelling in water should surely be properly rust-proofed. Trippel thought otherwise, meaning that many sprung leaks and sank when corrosion took hold. Then there was the fundamental issue that it was no good either as a road vehicle or as a boat. On tarmac, the upright body made the handling perilous, while in water, the gasoline engine was prone to seizure through water ingress.

SPECIFICATIONS

TOP SPEED:	113km/h (70mph)
0–96KM/H (0–60MPH):	42.9secs
ENGINE TYPE:	in-line four
DISPLACEMENT:	1147cc (70ci)
WEIGHT:	1041kg (2315lb)
MILEAGE:	8.8l/100km (32mpg)

Left: ***The Amphicar was sold all over Europe – this flyer was given to dealers in Denmark, where its aquatic abilities may have been useful.***

You would expect an amphibious car to be well waterproofed, wouldn't you? But many an Amphicar failed when water got into the engine, causing it to seize, while water traps in the body seams encouraged rust.

Sadly, the amphibious car was too much of a compromise. It wasn't brilliant in water, and it was awful on the road, offering terrible handling and no grip whatsoever, while the Triumph-sourced engine felt underpowered.

It was nothing if not innovative – to turn the Amphicar into a boat, all you had to do was drive it into the water, whereupon a propeller took over from the road wheels to provide motion.

ASIA ROCSTA *(1993–95)*

There have been several Jeep clones over the years, but one of the least subtle copies was the Asia Rocsta, introduced by Korean company Kia in 1986. The Rocsta was originally built for South Korea's military forces, and by the early 1990s was introduced into certain markets as a leisure vehicle, capitalizing on the growth of the potent SUV market in Europe. But while it looked fairly cool, the Rocsta was truly awful. It came with a choice of either a 1.8-litre pushrod gasoline engine or a smoky old Peugeot-sourced diesel, neither of which offered any performance.

The ride was harsh, the steering was vague and build quality was diabolical. Buyers might have overlooked this if the car had been an established brand, but the Asia Motors name was new to European markets and meant nothing to potential customers.

SPECIFICATIONS

TOP SPEED:	113km/h (70mph)
0–96KM/H (0–60MPH):	36.5secs
ENGINE TYPE:	in-line four
DISPLACEMENT:	1789cc (105ci)
WEIGHT:	3748lb (169ci)
MILEAGE:	11.3l/100km (25mpg)

Left: ***Considering the car's poor handling and bumpy ride, Asia Motors were correct to claim the Rocsta was not for the straight and narrow!***

Handling was terrible: the live rear axle and leaf spring combination meant the Rocsta was prone to rolling over, while the vehicle's uncompromising ride jarred occupants' spines.
The interior build quality was perhaps the Rocsta's biggest fault. The plastics were awful, the driving position uncomfortable and the column stalks used to snap off.
Power – if you could call it that – came from a choice of outdated Mitsubishi gasoline or Peugeot diesel engines.
N444 RCS

BOND 3-WHEELER (1948–70)

In Britain after World War II, there were legions of buyers who refused to take the newly introduced driving test. Instead, they took advantage of a loophole in the law which allowed them to drive a three-wheeled car on a motorcycle licence, for which there was no exam. Lancashire engineer Laurie Bond catered for their needs with a range of three-wheelers, the most distinctive being the Minicar. Built to seat two people, assuming they were very close friends, the Minicar was made out of plastic and had a motorcycle engine. It handled badly and the steering was bizarre: if you turned the wheel too far, the Bond would turn on its own axis! Horrid to drive, and a pain to motorists unfortunate enough to be stuck behind one, it was nonetheless a sales success.

SPECIFICATIONS

TOP SPEED:	56km/h (35mph)
0–96KM/H (0–60MPH):	not possible
ENGINE TYPE:	single-cylinder
DISPLACEMENT:	122cc (7ci)
WEIGHT:	no figure available
MILEAGE:	no figure available

Left: ***There are four people in this Bond Minicar. One can only assume that the children don't have any legs.***

The use of a fibreglass body tub meant the Bond Minicar didn't rust on the outside, but the steel backbone chassis became weak and collapsed, causing the stressed body to crack.
Creature comforts were at a bare minimum – there wasn't even a fuel gauge, and the space was badly cramped.
MARK
The steering was linked directly to the single front wheel, meaning you could turn it through 180°. It was possible to turn a Bond round entirely on its own axis.

CITROËN BIJOU *(1959–64)*

Citroën's inspiration for the Bijou came from admiring the work of British manufacturers, who were building special bodies on the chassis of conventional family cars. It decided to do the same with the 2CV, and the Bijou was built at the company's UK plant in Slough, Berkshire. It was styled by Peter Kirwin-Taylor, who had designed the beautiful Lotus Elite – but the Bijou certainly wasn't his greatest effort.

The styling was awkward and ugly, while the fibreglass bodywork was heavier than that of a standard 2CV, making the Bijou a heavy and unwieldy car with a top speed that couldn't quite top 80km/h (50mph). It was also more expensive than a Mini, which was much more fun to drive and far more practical. Unsurprisingly, nobody wanted a Bijou, and Citroën only managed to shift 207 examples in five years …

SPECIFICATIONS

TOP SPEED:	(73km/h) 45mph
0–96KM/H (0–60MPH):	not possible
ENGINE TYPE:	flat-twin
DISPLACEMENT:	425cc (26ci)
WEIGHT:	580kg (1288lb)
MILEAGE:	6.3l/100km (45mpg)

Left: *The Citroën Bijou was a uniquely poor seller, failing to compete with cars in a similar class, or cash in on its precursor, the 2CV.*

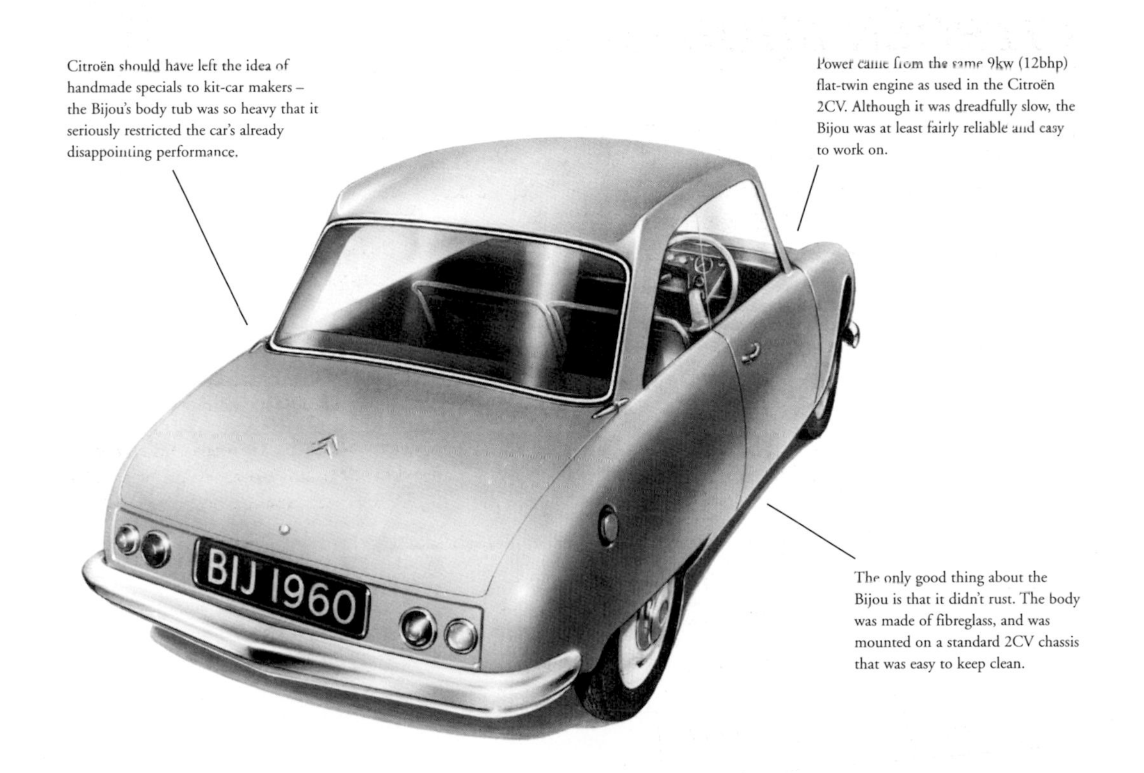

Citroën should have left the idea of handmade specials to kit-car makers – the Bijou's body tub was so heavy that it seriously restricted the car's already disappointing performance.

Power came from the same 9kw (12bhp) flat-twin engine as used in the Citroën 2CV. Although it was dreadfully slow, the Bijou was at least fairly reliable and easy to work on.

The only good thing about the Bijou is that it didn't rust. The body was made of fibreglass, and was mounted on a standard 2CV chassis that was easy to keep clean.

DACIA DUSTER (1983–90)

Another cynical attempt at trying to cash in on the growing SUV market, the Dacia Duster was actually a Romanian military vehicle, known in its home market as the ARO. It was simply constructed from corrugated panels, came with a choice of soft-top or metal roofs, and was based on a simple ladder-frame chassis. But while it had the off-road looks, the Duster didn't have the go-anywhere ability.

Power came from an ancient 1.4-litre Renault engine, offering dreadful performance in such a big vehicle, and it was primarily front-wheel drive. The rear axle could only be temporarily engaged by means of a special clutch. Buyers weren't impressed, particularly because Japanese rivals did it all so much better, and the Dacia Duster was quickly forgotten.

SPECIFICATIONS

TOP SPEED:	116km/h (72mph)
0–96KM/H (0–60MPH):	22.7secs
ENGINE TYPE:	in-line four
DISPLACEMENT:	1397cc (85ci)
WEIGHT:	1208kg (2685lb)
MILEAGE:	1.0l/100km (28mpg)

Left: ***This is Duster action! The car is off-roading to the best of its abilities – on a patch of wet grass.***

Who were they trying to kid? Despite its off-road pretensions, the Duster was based on the platform of the Renault 12 and offered little more in the way of ground clearance. That meant its off-road ability was limited.

There was nothing luxurious about the Duster's cabin. The vinyl dashboard was basic and the seats were uncomfortable, while space in the back was especially cramped.

It was a proper four-wheel drive, but the system was very crudely assembled. The second axle was engaged by means of a special clutch that linked the front and rear wheels together.

DATSUN CEDRIC/300C (1961–80)

In order to strengthen sales in export markets, Datsun (later Nissan) decided to introduce a large executive car to appeal primarily to wealthy male buyers. So it needed a suitably masculine name. Cedric wasn't it … Inspiration came from the story *The Little Prince*, in which the name of the hero was Cedric. Datsun wanted to give the car a 'distinguished English-sounding name' that would appeal to upper-class customers. Instead, they ended up with an old-fashioned and wimpish moniker. In 1966, the 'Cedric' was dropped in export markets, as it was felt that the unfortunate name was hindering sales. Nor were these helped when Nissan later changed the name to 'Gloria', another disaster. Worse, this was not even an especially attractive car. Incidentally, Nissan Cedric is an anagram of 'canned crisis'. Strangely apt …

SPECIFICATIONS

TOP SPEED:	200km/h (120mph)
0–96KM/H (0–60MPH):	9.9secs
ENGINE TYPE:	in-line six
DISPLACEMENT:	2393cc (146ci)
WEIGHT:	1232kg (2737lb)
MILEAGE:	12.8l/100km (22mpg)

Left: ***Check out this seat of pleated velour and grey plastics, then ask yourself this: Wouldn't you rather have a Mercedes-Benz?***

The Cedric always had a decent engine, with a choice of six-cylinder units. Later 300C examples had a 3-litre (183ci), straight-six and were capable of cruising at over 193km/h (120mph), but the soft suspension and vague steering meant it was never sporty.

There were three Datsuns that wore the Cedric name. This was the final example, which lasted from 1977 to 1980.

Japanese makers have never quite got the measure of luxury car interiors, and the Cedric was no exception. It had equipment to rival the best European manufacturers, but the design was bland and the finish cheap.

DUTTON SIERRA *(1977–91)*

Perhaps the most interesting thing about the Dutton Sierra was that it spawned a remarkable David-versus-Goliath court case, when global giant Ford tried to sue the tiny British kit car maker for using the name of its family saloon. But Ford didn't prepare its case properly, and ended up having to compensate Dutton, which had used the name first. Apart from that, the Sierra was spectacularly unimpressive. It had the looks of an off-roader, but it was based, ironically, on a standard rear-drive Ford platform, and its fibreglass panels very rarely fitted properly.

Driving dynamics weren't helped by its tall bodywork, which made it feel decidedly top-heavy, while in styling terms it was gawky and unpleasant. Not an impressive car, but one with an interesting story to tell …

SPECIFICATIONS

TOP SPEED:	145km/h (90mph)
0–96KM/H (0–60MPH):	no figures available
ENGINE TYPE:	in-line four
DISPLACEMENT:	1596cc (92ci)
WEIGHT:	no figures available
MILEAGE:	no figures available

Left: ***Inside the Sierra, it was a case of spot the parts source. The Sierra's dash was a mixture of other manufacturer's old bits.***

Fibreglass panels didn't rust, but they were never quite cut properly, resulting in shocking panel gaps and doors that were always difficult to close.

Despite the raised ground clearance, created by using springs from Ford Transit, the Dutton Sierra was based entirely on the platform of a Mk 2 Ford Escort station wagon, right down to the rear cart springs.

It was ironic that Ford tried to sue Dutton, as the Sierra's mechanical layout was made almost entirely of Ford bits, most of which were taken from the Mk 2 Escort and Mk 4 Cortina.

HYUNDAI ATOZ *(1997–2000)*

If you were looking for the ultimate in ugliness and uninspiring driving in the late 1990s, the Hyundai Atoz was it. Sold elsewhere as the Atos, it was changed for English-speaking markets to Atoz, lest people felt they couldn't give a … And you'd be forgiven for not doing so, as the Atoz was a truly awful machine. It may not have fallen out of the ugly tree, but it had certainly crashed into it at high speed, with angular styling that had no harmony.

The ungainly appearance could have been forgiven if the Atoz had been spacious and comfortable inside, but Hyundai had sacrificed some of the car's practicality to make the styling possible. The trunk was practically useless, the doors were too narrow and the cabin was cramped. In a market where small cars were all about space-creating designs, it had no place whatsoever.

SPECIFICATIONS

TOP SPEED:	152km/h (94mph)
0–96KM/H (0–60MPH):	15.4secs
ENGINE TYPE:	in-line four
DISPLACEMENT:	1086cc (70ci)
WEIGHT:	885kg (1967lb)
MILEAGE:	7.0l/100km (40mpg)

Left: ***This family is clearly enjoying their Atoz, despite the fact that the trunk is only big enough to hold a basket of fruit.***

The stylist made the front look equally incongruous. The round headlights were at odds with the car's boxlike styling, and the grille looked like rabbits' teeth.

Cheap plastics, a tiny trunk and an awkward driving position meant that the Atoz wasn't even a competent small car – it was compromised in far too many respects and it received several deservedly critical reviews as a result.

An object lesson in how not to style a small car: around the licence plate, the Atoz's rear end featured a chrome strip that looked like a droopy moustache.

LADA NIVA *(1979–96)*

Show a Lada Niva a tough off-road course, and it wallows around like a happy pig, leaping over mud banks, wading through stagnant water and slithering its way up filthy tracks as if it was born to roll around in muck. But show it a twisty stretch of tarmac, and you'll learn it handles like a pig as well, with excessive body roll, a complete lack of steering accuracy and an appalling, spine-jarring ride. Engines came from the Lada Riva and were coarse and thirsty, build quality was atrocious, the black plastic interior was sombre and the brakes were all too literally a hit-or-miss affair.

Yet the Niva had a strong following with European farmers, as it was sold at a bargain price and matched a Land Rover off-road for a quarter of the price. For others, though, it was a dreadful car, as many initially optimistic buyers soon found out …

SPECIFICATIONS

TOP SPEED:	124km/h (77mph)
0–96KM/H (0–60MPH):	22.4secs
ENGINE TYPE:	in-line four
DISPLACEMENT:	1569cc (91ci)
WEIGHT:	2604lb (1172kg)
MILEAGE:	28mph (10.0l/100km)

Left: ***Now that's style! Niva owners could support the Lada cause with some specially branded clothing. We'll take the bomber jacket, please …***

The Niva is nothing if not robust. The all-round coil springs, mounted on a ladder frame chassis, gave it amazing axle articulation, and it was brilliant off-road. Sadly, it was appalling to drive on tarmac, with vague handling and a bouncy ride.

Cheapness is the hallmark of the Niva's cabin – it's badly laid out, with lots of nasty plastics and horrible vinyl seat facings.

LIGIER AMBRA *(1998–present)*

There's a strong market in France for small four-wheeled vehicles with an engine capacity of less than 500cc. The Ambra is one such car, as a dispensation in the law allows them to be driven by people who don't have driving licences, often from as young as 14 years of age. That's a terrifying prospect when you consider just how difficult the Ambra is to drive. The cabin is impossible to sit comfortably in, the brakes do a very poor impression of actually working and the handling is terrifying. Approach the same corner several times at the same speed, and each time the Ambra will handle differently. A truly, truly awful motor car that is deeply unpleasant in almost every respect.

SPECIFICATIONS

TOP SPEED:	100km/h (62mph)
0–96KM/H (0–60MPH):	no figures available
ENGINE TYPE:	flat-twin
DISPLACEMENT:	505cc (31ci)
WEIGHT:	no figure available
MILEAGE:	4.7l/100km (60mpg)

Left: ***According to the advertisement, the Ambra came in both 'Country' and luxury versions, both of them horrible to drive.***

You need to be fairly athletic to get comfortable in the Ambra. The driving position is terrible, with the pedals offset to the right, and a lack of seat adjustment makes it impossible for tall people to get comfortable inside.

You know the Ambra is a bad car as soon as you switch it on. The engine is crude and noisy, and there's hardly any soundproofing. Acceleration is also dreadfully slow.

It's just as well that the Ambra can't go very quickly because the handling is appalling. It has no front-end grip, the steering lacks any kind of feel and the basic suspension set-up causes it to lurch and wallow round bends.

LOTUS SEVEN S4 *(1970–73)*

Car designers had some funny ideas in the 1970s, and the 1970 Lotus Seven S4 is a classic example. For some reason, the British firm thought it could improve on the timeless original Seven – which today remains in production as a Caterham. To do so, it completely redesigned the body, replacing the original car's pretty alloy tub with a hideous fibreglass contraption that sat far too high and had sharp edges that sliced off the owners' skin if they weren't careful getting in and out.

The steering and suspension were also revised, making the car softer and less direct than before, but at least the performance was good. Sadly, though, this wasn't enough to revive sales, and Lotus only ever built about 1000 Seven S4s before finally canning the project in 1973.

SPECIFICATIONS	
TOP SPEED:	162km/h (100mph)
0–96KM/H (0–60MPH):	8.8secs
ENGINE TYPE:	in-line four
DISPLACEMENT:	1598cc (92ci)
WEIGHT:	1276lb (574kg)
MILEAGE:	30mpg (9.4l/100km)

Left: ***Not even the Lotus Seven could escape the excesses of the 1970s. Check out the styling kit and the horrible whitewall tyres!***

Slim people only, please! If you carried a bit of weight, the Seven S4's uncompromisingly basic cabin was very uncomfortable. The seats were so narrow that you could get yourself stuck between the transmission tunnel and doors.

The entire hood hinged up for easy access, and underneath it you'd find a choice of Ford Cortina-sourced engines. These were basic, but provided more than enough power for the S4's light weight.

You can't improve on perfection – a lesson that Lotus learned the hard way with the Seven S4. The original's Seven's simple yet beautiful lines were ruined by 1970s' excess.

MARCOS MANTIS *(1970–71)*

If prizes were awarded for ugliness, the Marcos Mantis would have a bulging trophy cabinet. Introduced in 1970, the Mantis had ridiculous styling. It was an incongruous mix of a wedge-shaped nose, square headlights and a curvaceous, swooping rear end, and fans of the company's earlier, pretty offerings found it vile by comparison. It was also a four-seater, which Marcos believed would attract a new type of buyer, as well pleasing its traditional followers.

But in truth, it didn't satisfy either camp. Its vulgarity meant it didn't attract enough new customers to the brand, while it also scared off several existing fans, who went for much more attractive TVRs instead. The Marcos company folded in 1971, after only 32 cars had been built …

SPECIFICATIONS

TOP SPEED:	203km/h (125mph)
0–96KM/H (0–60MPH):	8.4secs
ENGINE TYPE:	in-line six
DISPLACEMENT:	2498cc (152ci)
WEIGHT:	1035kg (2300lb)
MILEAGE:	14.8l/100km (19mpg)

Left: ***Here, a driver is sneaking through the forest in his Mantis, hoping that nobody will see him behind the wheel.***

The cabin has little going for it: after a raid on the parts bins of both British Leyland and Ford, the Mantis emerged with a haphazard arrangement of switchgear, mostly plucked from ordinary British saloon cars.

The Mantis was a styling travesty; it has to be one of the ugliest cars ever manufactured, and it's a surprise that the designers ever really signed it off for production.

The Mantis's looks could have been forgiven if it had been great to drive – but it wasn't. Performance was merely average, and the handling was far too tail-happy and difficult to control.

MASERATI BITURBO *(1981–91)*

Maserati tried to go more conventional than usual with the Biturbo, and introduced a car that was cheaper and more ordinary-looking than previous offerings. It was supposed to take on the likes of the BMW 635 and Jaguar XJS, but actually it looked stupid. For starters, there was the styling: boring, like an early 1980s American sedan rather than a sports coupé. The only saving grace was a beautifully ornate interior. That wasn't enough, though, as the Biturbo suffered from horrendous turbo lag.

It was slow off the mark, then the V6 engine's twin turbos could kick in with dramatic effect. Coupled to the rather crude chassis, this made it prone to terrifying oversteer, especially if the turbos came in mid-bend.

Left: *A self-proclaimed 'modern day classic' according to Maserati, but someone obviously forgot to tell the styling department.*

SPECIFICATIONS

TOP SPEED:	203km/h (126mph)
0–96KM/H (0–60MPH):	7.3secs
ENGINE TYPE:	V6
DISPLACEMENT:	2491cc (152ci)
WEIGHT:	1233kg (2739lb)
MILEAGE:	11.3l/100km (25mpg)

Maserati tried a less dramatic approach when styling the Biturbo, but instead of creating a car that was elegant in its simplicity the designers came up with a generic design that looked far too much like a Chrylser Le Baron.

Italian luxury opulence was very much in evidence, with beautiful hand-stitched leather seat facings, suede detailing and gold-edge dashboard dials.

Handling was terrifying: with appalling turbo lag and a crude suspension setup, the Biturbo couldn't cope with the power delivered by its V6 engine, and it was prone to spinning at the merest provocation.

MATRA RANCHO (1978–84)

Although it was a pretty poor car, there was a lot to like about the Matra Rancho's distinctive design. In many respects, it was ahead of its time. Following the unmitigated success of the Range Rover, Matra introduced a car that had all of the off-roader's butch looks, but without its expensive four-wheel-drive transmission. At the front, it was a Simca 1100, but the fibreglass rear section was entirely unique, and there was even a seven-seat option. It should have been an enormous success. But build quality was terrible, and, while the fibreglass rear section never looked rusted, the underlying metal was prone to excessive corrosion. Despite the looks, it had absolutely no off-road ability, and, although the concept of an off-road look is popular today, the Rancho never really made an impact on the marketplace in its heyday.

Left: ***The Canterbury Motor Company wanted £5400 (US$12,000) for a new Rancho – a lot of money in 1981, when this advert first appeared.***

SPECIFICATIONS

TOP SPEED:	144km/h (89mph)
0–96KM/H (0–60MPH):	14.9secs
ENGINE TYPE:	in-line four
DISPLACEMENT:	1442cc (88ci)
WEIGHT:	1179kg (2620lb)
MILEAGE:	8.8l/100km (32mpg)

When it first came out, many people thought the Rancho's 4x4 looks and conventional underpinnings were a bit of a joke. But with hindsight, it was actually a precursor to a whole new generation of 'soft-roaders'.

Practicality was one of the Rancho's biggest plus points. The trunk was enormous and it had a Range Rover-style split tailgate, and there was even a fold-flat rear seat that could be stowed in the floor.

It was impossible to tell whether or not a Rancho was starting to rust. The fibreglass rear end could conceal all manner of structural corrosion problems on the steel floor underneath it.

MG MONTEGO *(1984–91)*

If at first you don't succeed, try again … That was Austin-Rover's philosophy, as the MG version of the Maestro had developed into a sales and marketing disaster. The Montego was better, but it was still a very cynical interpretation of the classic MG name, and traditional fans of the marque were deeply offended by what had become of their beloved car company. The MG Montego used exactly the same 2.0-litre fuel-injected engine as an ordinary 2.0-litre Montego and was no more dynamic to drive, with only a digital dashboard and red seatbelts marking it out as anything different. Its image wasn't helped, either, by the alarmingly premature body rot problems suffered by all Montegos.

The Turbo model, which was launched in 1985, at least had some interest, although it suffered from a disconcerting and curious mix of turbo lag and torque steer.

SPECIFICATIONS

TOP SPEED:	184km/h (114mph)
0–96KM/H (0–60MPH):	9.6secs
ENGINE TYPE:	in-line four
DISPLACEMENT:	1994cc (122ci)
WEIGHT:	1022kg (2271lb)
MILEAGE:	10.0l/100km (28mpg)

Left: ***Black and red were essential if you wanted to make something appear sporty in the 1980s. But the Montego's tawdry styling was never going to impress.***

Austin-Rover fitted a turbocharger to the MG Montego, which made handling very interesting indeed. When the turbo kicked in, the front wheel would scrabble for grip and threaten to rip the steering wheel out of the driver's hands.

Digital dashboards were all the rage in the 1980s, and the MG Montego's was particularly rooted in the era. The LCD rev counter would start off green, turn yellow as the revs increased, then flash red when it was time to change gear.

Standard Montegos were awful rust buckets, and the MG versions were no better – and the plastic wheel arch covers disguised the corrosion until it had become terminal.

NASH METROPOLITAN (1953–61)

Built in cooperation with British manufacturer Austin, the Metropolitan was the vision of Nash president George Mason. It was planned to be a small car that would combine the styling to which US buyers were accustomed to with the low running costs of a small model, and Austin's success with compact cars was legendary. Sadly, it was a disaster.

The styling looked daft, and the high-sided body was prone to rusting. Add basic 1950s Austin underpinnings, and you get handling that was at best erratic, and at worst downright dangerous.

The brakes were poor, too, requiring plenty of forward planning and providing little in the way of emergency stopping power. And it wasn't even practical – despite the appearance, it didn't come with an opening trunk lid, and the car's cabin was cramped.

Left: ***The logic behind the Nash Metropolitan raises many questions, but the one we want answered is this: Why on earth is there a ballerina on the beach with a pair of water skis?***

SPECIFICATIONS

TOP SPEED:	121km/h (75mph)
0–96KM/H (0–60MPH):	24.6secs
ENGINE TYPE:	in-line four
DISPLACEMENT:	1489cc (91ci)
WEIGHT:	945kg (2100lb)
MILEAGE:	8.6l/100km (33mpg)

The Metropolitan used BMC's proven and reliable B-Series engine, which gave it reasonably good performance, but the chassis and brakes simply couldn't cope, and the car needed to be driven with caution.

Although it looked like a conventional shape, the Metropolitan didn't get itself an opening trunk until it had received a mid life facelift. Prior to that, the only way to access the luggage hold was to drop the back seat.

Underneath, the Nash Metropolitan was based on the platform of the Austin A40. This was deeply unsporting, with heavy steering, a choppy ride and unnerving handling characteristics.

NISSAN SUNNY ZX (1985–90)

With compact performance cars undergoing something of a renaissance in the mid-1980s, Nissan decided it had to get in on the act. The Sunny ZX was its response to such revered rivals as the Peugeot 205 GTI 1.9 and Volkswagen Golf GTI 16v – both seriously capable opponents. Undeterred, Nissan shoehorned in a respectably lively 1.8-litre twin cam engine, which was by far the ZX's greatest asset. In all other respects, it was a horrible machine.

The tacked-on spoilers and side skirts were a poor fit and did nothing for the Sunny Coupé's already boxy styling, while the alloy wheels were ostentatious, but without any style. The ride quality was so harsh you could feel every jolt through the seats, and, while the handling was acceptably assured, it wasn't a patch on the ZX's closely priced competitors. Pointless, and the sales figures proved it.

SPECIFICATIONS

TOP SPEED:	206km/h (128mph)
0–96KM/H (0–60MPH):	8.3secs
ENGINE TYPE:	in-line four
DISPLACEMENT:	1803cc (110ci)
WEIGHT:	1364kg (3031lb)
MILEAGE:	8.8l/100km (33mpg)

Left: *How do you indulge yourself and stay in good shape? Well, we're not sure, but we doubt it has anything to do with the Nissan Sunny ZX.*

The Sunny ZX wasn't attractive to start with, but when Nissan added a body kit to accentuate its trapezoid angles, it managed to look even more daft.

In typically Japanese fashion, the Sunny ZX came with an exceedingly bland interior layout. The dashboard was coated in nasty plastics, and the dials and switches were all over the place.

The ZX was surprisingly quick – its 1.8-litre (110ci) engine soon found popularity with smaller sports car makers, who used it under licence.

OLDSMOBILE TORONADO (1966–70)

Front-wheel drive might not be the most obvious or appropriate way of powering a high-performance supercar, but that's exactly the route General Motors followed with the Oldsmobile Toronado. Ironically, it worked quite well, giving greater agility and grip than most rear-drive rivals. But unfortunately for GM, American buyers hated the idea of front-drive in such a powerful car and voted with their wallets, buying other brands and leaving the unusually styled Toronado as a multi-million dollar investment that never got the returns it deserved. It also suffered from excessive front-tyre wear, very poor fuel consumption and inefficient drum brakes. An adventurous and technologically advanced model, but not a success.

SPECIFICATIONS

TOP SPEED:	208km/h (129mph)
0–96KM/H (0–60MPH):	8.1secs
ENGINE TYPE:	V8
DISPLACEMENT:	6965cc (425ci)
WEIGHT:	2102kg (4670lb)
MILEAGE:	18.8l/100km (15mpg)

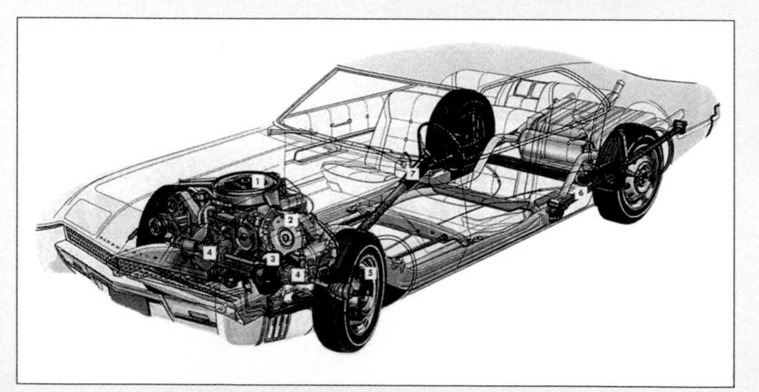

Left: ***This cutaway drawing shows the complex engineering layout of the Toronado, with its longitudinally mounted engine and front-wheel drive.***

Performance was astounding – it came with a 7.4-litre (454ci) V8 engine, and was capable of topping 210km/h (130mph).

Front-wheel drive was entirely new to American car makers, and, while the Toronado had suitably assured handling for such a big car, it failed to find favour with buyers, who expected performance cars to have power at the rear.

The Toronado was enormous, and its styling was certainly uncompromising. But in later years, the huge panels suffered from corrosion, and the doors drooped on their hinges thanks to rot in the A-pillars.

PANTHER J72 *(1972–81)*

You could have bought an infinitely better V12 Jaguar E-Type for the same money as a Panther J72 – and that was just one of its many problems. Built as a plaything for ostentatious, well-heeled customers, it mimicked the look of the original Jaguar SS100, but with several tacky and tasteless 1970s adornments such as chrome alloy wheels, bucket seats and vinyl interior trim. Although powered by gutsy Jaguar engines, the Panther J72's performance was stunted by terrible aerodynamics, which made the front end prone to lifting and drifting at high speed.

On top of that, the steering was directly linked and unpleasantly heavy to use, power delivery was brutally unrefined and the ride was poor. It was well screwed together, but ultimately lacked any kind of finesse in its finish. An especially rare vulgarity – and deservedly so …

SPECIFICATIONS

TOP SPEED:	219km/h (136mph)
0–96KM/H (0–60MPH):	no figures available
ENGINE TYPE:	V12
DISPLACEMENT:	5343cc (326ci)
WEIGHT:	1127kg (2504lb)
MILEAGE:	18.8l/100km (15mpg)

Left: ***In our book, classic automobiles of the 1980s are icons such as the Audi Quattro and Peugeot 205 GTI – and not revolting pastiches of 1930s' greats. Sorry, Panther.***

At least the power plants were genuine Jaguar. The entry-level car came with the straight-six unit from the XJ6 saloon, while more powerful examples got the V12 unit from the XJ-S and E-Type.
Trying to copy the Jaguar SS100 was considered an affront by purists, especially as the Panther J72 was completely over the top in its styling. Many critics considered it tasteless.
The J72 was crudely assembled, and was far too powerful for its chassis. It was excessively tail-happy, and proved challenging to drive.

RELIANT ROBIN/KITTEN (1974–89)

British fibreglass specialists Reliant had an unusual model range to say the least. At the top was the dramatic and imposing Scimitar sports coupé, there was no middle, and at the bottom was the lowly Robin. Possibly the world's most famous three-wheeled vehicle, the tripod of the tarmac was nonetheless a pretty awful car. It had a lively four-cylinder, all-alloy engine, but it was impossible to enjoy it to the full because attacking any corner at speed resulted in the Robin falling over.

As a more stable alternative, Reliant offered the four-wheeled Kitten, which was in all other respects the same car – but, unlike the Robin, you couldn't drive it with just a motorbike licence, and it was priced to compete with other, much more competent small cars.

SPECIFICATIONS

TOP SPEED:	126km/h (78mph)
0–96KM/H (0–60MPH):	19.6secs
ENGINE TYPE:	in-line four
DISPLACEMENT:	848cc (52ci)
WEIGHT:	522kg (1159lbs)
MILEAGE:	7.0l/100km (40mpg)

Left: ***The Kitten was the car built to stay young, but unfortunately the average buyer was well into their sixties.***

Power came from an all-alloy engine designed and built exclusively by Reliant. It was quite lively given its compact displacement.

The four-wheel Kitten was surprisingly agile to drive, with handling to rival a Mini. The three-wheeled Robin, on the other hand, was dangerous if cornered at any speed, as the car was all too easy to tip over.

Fibreglass bodies were durable and kept rust at bay – the chassis underneath was so simple that repairs were inexpensive and easy, meaning many Robins and Kittens had a long lifespan.

SEAT 1200 SPORT *(1972–86)*

Spanish firm SEAT had been making Fiats under licence for almost two decades when it finally got the approval to design and build its own car. The 1200 Sport used the floorpan, engine and transmission of a Fiat 128 to keep build costs to a minimum, but the exterior and interior were entirely SEAT's own design. Sadly, they were so badly executed that, in one fell swoop, one of the ugliest sports cars ever was created. The boxy coupé lines wouldn't have been so bad, were it not for the fact that the 1200 Sport was then plagued with an unsightly black polyurethane nose, which looked like an amateur afterthought. Things didn't get much better inside, with an expanse of black plastic and trim that didn't so much fall of the dash as leap off it with glee …

SPECIFICATIONS

TOP SPEED:	150km/h (93mph)
0–96KM/H (0–60MPH):	no figures available
ENGINE TYPE:	in-line four
DISPLACEMENT:	1116cc (68ci)
WEIGHT:	804kg (1787lb)
MILEAGE:	8.8l/100km (32mpg)

Left: ***With acres of brown velour and black vinyl, plus that sexy two-spoke steering wheel, the 1200 Sport oozed Hispanic machismo. Until it fell apart.***

The engine came directly from Fiat, and was the same unit that powered the 128 saloon. Despite the 1200 Sport's name, however, it was never a performance car.

SEAT's first exclusively styled car was a brave move – but not an entirely successful one. It was done on the cheap, and the polyurethane nose cone was prone to warping and coming loose in hot temperatures.

The platform was also Fiat's, and offered reasonable road-holding and tidy handling. The 1200 Sport could have been a good car had a bit more attention to detail been applied to its design.

SUBARU XT COUPÉ (1985–90)

Long before the all-conquering rally cars arrived, Subaru tried to get sporty, and the car with which it hoped to boost its performance image was this: the XT Coupé. Not only was it quite obviously a styling aberration of the highest order, but it was also crudely engineered. It was based on the platform of the standard 1800 saloon, which in turn formed the basis of the Subaru pick-up so beloved of farmers across the globe. That meant it was fairly agricultural to drive, although the four-wheel drive on offer at least meant it handled well despite the bouncy ride.

Still, it was powerful as long as you didn't mind waiting for the turbo to spool up, meaning you could usually make a quick getaway before anyone looked inside and noticed the futuristic all-plastic dashboard – an exercise in appalling taste if ever there was one.

SPECIFICATIONS

TOP SPEED:	190km/h (118mph)
0–96KM/H (0–60MPH):	9.5secs
ENGINE TYPE:	flat-four
DISPLACEMENT:	1781cc (110ci)
WEIGHT:	1135kg (2523lb)
MILEAGE:	10.9l/100km (26mpg)

Left: ***It's no wonder this farmer is angry with his son. He was obviously expecting him to come home with a useful pick-up, not this styling travesty.***

Subaru was aiming for a youthful and sporty demographic with the XT, but it misread the market. The digital dials and tacky plastics were considered deeply uncool.

Wedge-shaped cars went out of fashion in the late 1970s, but somebody forgot to tell Subaru. The XT Coupe looked like a Triumph TR7 on stilts …

It's a Subaru, which means that the XT has a flat-four engine and permanent four-wheel drive. That makes it great to drive, but it's so ugly nobody cares …

SUZUKI X90 (1997–99)

The X90 was nothing if not brave – with the global car market for ever branching out into more and more unusual niches, Suzuki hit on an unexplored market area all of its own when it introduced its new four-wheel-drive model in 1997. The X90 was the world's first – and only – two-seater, sports car off-roader. Or at least that's what it was meant to be.

In truth, it was nothing more than an old, agricultural and largely unpleasant Vitara, with an underpowered 1.6-litre engine and all of the model's practicality removed to scale down the cabin and replace it with a bubble that looked vaguely like an aircraft cockpit. The X90 was neither fast, sporty nor sensible as an off-roader; to put it more bluntly, it was useless. Why on earth did they bother?

SPECIFICATIONS

TOP SPEED:	179km/h (111mph)
0–96KM/H (0–60MPH):	10.5secs
ENGINE TYPE:	in-line four
DISPLACEMENT:	1590cc (92ci)
WEIGHT:	981kg (2180lb)
MILEAGE:	9.4l/100km (30mpg)

Left: ***Suzuki tried its best to hide the X90's styling by showing only the tail light on the brochure cover. The true horrors were revealed when you turned the page.***

Was it a 4x4, or a sports coupé? While the roof suggested the latter, the rest of the X90's styling pointed firmly to the car's Suzuki Vitara origins …

Power came from a standard Vitara engine, meaning it wasn't especially quick, nor was it refined or economical. Few buyers were impressed, and the X90 was a giant flop.

Because of its agricultural origins, the X90 failed to deliver any kind of dynamic thrills. It also had dire ride comfort, with the suspension thumping and crashing over uneven surfaces.

TOYOTA CROWN *(1967–74)*

When Toyota started to get its export markets into shape in the early 1970s, its success was based on a range of affordable, well-built but ultimately nondescript saloons and coupés. There was, however, one notable exception – and it was certainly distinctive. The Series 2 Crown was introduced in 1971, and it was a perfect example of just how badly Japanese manufacturers misunderstood the luxury car market. The Crown was built on a sound basis – it was large, spacious and well equipped, and had automatic transmission and high performance (162km/h).

But Toyota overlooked one fundamental element, and that was the styling. Having heard that European and American buyers liked their cars to have individualistic styling, Toyota went to town on the Crown – and in the process created what was one of the most unharmonious, repulsive shapes ever to disgrace our roads …

SPECIFICATIONS

TOP SPEED:	162km/h (100mph)
0–96KM/H (0–60MPH):	12.7secs
ENGINE TYPE:	in-line six
DISPLACEMENT:	2563cc (154ci)
WEIGHT:	1301kg (2890lb)
MILEAGE:	12.8l/100km (22mpg)

Left: ***Styling excess was what the Crown was all about; Toyota completely misjudged the requirements of the executive car market.***

Toyota was desperate to make the Crown appeal to executive buyers, but it went over the top with styling. The car's chrome-encased nose and twin headlights were especially unpleasant.

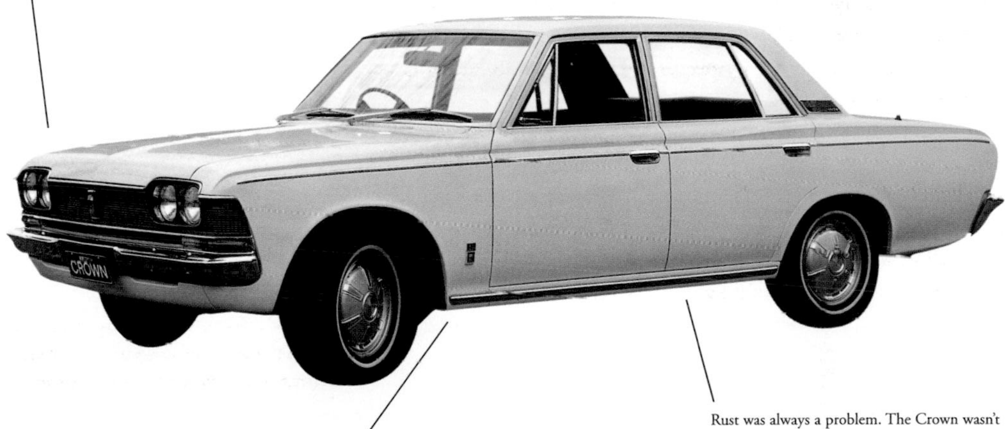

One thing a luxury car needs is an elegant interior, but somebody forgot to tell Toyota. The Crown was trimmed in velour, and used the same shiny black plastics as lesser models in the Japanese firm's range.

Rust was always a problem. The Crown wasn't brilliantly well made, and rust found its way into the sills, doors, inner wings and trunk floor with alarming speed.

TRABANT 501 *(1965–91)*

There was only one time in its 30-year life that the Trabant found fame – and it was one of the most significant moments of the twentieth century. When the Berlin Wall fell in 1989 and Germany was reunited, newsreels were packed with traffic jams full of the former East German model chugging through the Brandenberg gate. It was, of course, a car that that gave freedom to many people: it mobilized the impoverished and was essential to the people that owned it, meaning that it's also one of the most loved cars ever made.

But good? No. The Trabant is truly awful. The body is made of an odd, compacted cardboard material known as Duraplast, the two-stroke engines belch out filthy smoke, performance is dreadful and the gear change is among the worst in the world. An awful car – but not without its charms.

SPECIFICATIONS

TOP SPEED:	100km/h (62mph)
0–96KM/H (0–60MPH):	no figure available
ENGINE TYPE:	flat-twin
DISPLACEMENT:	594cc (36ci)
WEIGHT:	no figure available
MILEAGE:	9.7l/100km (29mpg)

Left: ***'Ich Fahre Einen Trabant' simply means 'I drive a Trabant'. As did everyone else in the former East German Republic …***

Trabants were made out of an unusual material called 'Duraplast'. This was extremely cheap, as it consisted mainly of compressed cardboard, coated in a plastic glaze. Crashing one, however, was not recommended!

Even when new, a Trabant belched out a trail of smoke. The two-stroke engine was designed with mechanical simplicity in mind, not exhaust emissions …

In terms of layout, the Trabant was well designed. The cabin had ample seating for four, and it was surprisingly comfortable, if a little basic.

TRIUMPH MAYFLOWER (1949–53)

With post-war austerity still hanging over Britain, Standard Triumph recognized a gap in the market for a model that had all the trappings of a luxury car, but which was clothed in a compact body and equipped with a low capacity, economical four-cylinder engine. The Triumph Mayflower could have been a huge success, but Triumph took the traditional styling a little too far, giving it the appearance of a Rolls-Royce Phantom that had been chopped in the middle. The razor-edge upper styling and curved lower panels looked decidedly stupid when mated together and, to make matters worse, the crude chassis made for perilous handling, along with a fairly wayward steering setup and brakes that were little more than a token gesture. The Mayflower was dropped after just four years, even though it had been reasonably popular with middle-class buyers.

SPECIFICATIONS

TOP SPEED:	101km/h (63mph)
0–96KM/H (0–60MPH):	no figure available
ENGINE TYPE:	in-line four
DISPLACEMENT:	1247cc (76ci)
WEIGHT:	907kg (2016lb)
MILEAGE:	7.2l/100km (39mpg)

Left: ***Advertising in the UK during the 1950s was always fairly reserved and lowbrow. 'Britain's New Light Car' was hardly the snappiest of slogans to draw prospective buyers into the showrooms.***

Under the hood, the Mayflower came with a pre-war side-valve engine, which offered dismal performance. It was also coupled to an unpleasant column-change gearbox.

The Mayflower was designed to look like a miniature Rolls-Royce Silver Dawn, so two-tone paint and a noticeable 'waistline' in the body were de rigeur. But it was so small, it just looked pretentious and daft.

Handling was perilous. The car's simple construction and crude suspension, along with the upright body, meant that it leaned badly in corners and lurched out of control if driven too quickly through a bend.

WARTBURG 353 KNIGHT *(1966–88)*

If ever there were an award for the most appalling car ever built, the Wartburg Knight would have stood a pretty good chance of winning. The East German saloon lasted for 24 years and sold well over a million, despite being fundamentally awful. It was front-wheel drive, but was horrendous to drive, with frightening handling and no front-end grip whatsoever. On top of that, it was powered by a smoke-belching, two-stroke engine, was prone to excessive corrosion and had various bits of trim that parted company with the car at random moments, often with the bolt-on panels attached. It had dreadful performance, struggling to top 110km/h (70mph), yet in return gave only 10l/100km if driven carefully. For the ultimate in crude and unpleasant transport, there was even a station wagon model called the Tourist.

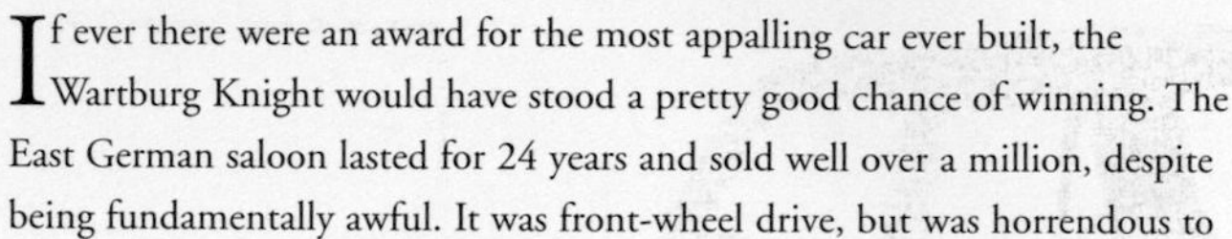

Left: *The best the advertising people could come up with was that the Wartburg Knight was a 'full size estate car for £750'. Not an especially good one, then?*

SPECIFICATIONS

TOP SPEED:	119km/h (74mph)
0–96KM/H (0–60MPH):	22.8secs
ENGINE TYPE:	three-cylinder two-stroke
DISPLACEMENT:	991cc (60ci)
WEIGHT:	878kg (1952lb)
MILEAGE:	0.9l/100km (26mpg)

The Wartburg was one of the most polluting cars ever built – it's three-cylinder, two-stroke engine belched out tons of smoke and drank oil.

For ease of repair and maintenance, the Wartburg's rear body panels were bolted in place, rather than welded. This made repairs easier, and on station wagon models the whole rear section was moulded out of fibreglass to make it simpler to fit in the factory.

Although the Wartburg's tail-happy characteristics made it quite successful in rallies, the unpredictable handling was too much for the European buying public. Several car magazines decried it as dangerous.

Index

Page numbers in *italics* refer to illustrations.

Picture credits

Art-Tech/Midsummer Books: 11, 13, 25, 41, 47, 59, 63, 65, 89, 91, 93, 113, 131, 147, 151, 155, 157, 167, 171, 173, 175, 179, 183, 187, 228, 229, 263, 287, 289, 295, 299;

Cody Images: 55, 153, 247, 309;

Giles Chapman Library: 33, 39, 45, 70, 99, 105, 111, 121, 122, 123, 124, 125, 139, 142, 149, 159, 163, 196, 209, 211, 213, 215, 221, 227, 233, 237, 243, 245, 265, 268, 275, 279, 297, 307;

Craig Cheetham: 109, 253, 303, 313;

The Culture Archive: 28, 29;

Richard Dredge: 8, 10, 12, 14, 15, 18, 19, 20, 22, 23, 24, 26, 27, 30, 31, 32, 34, 35, 36, 37, 38, 40, 42, 43, 44, 46, 48, 49, 50, 51, 52, 54, 56, 58, 60, 61, 62, 64, 66, 67, 68, 72, 74, 75, 76, 77, 78, 79, 80, 81, 86, 87, 88, 90, 94, 95, 96, 98, 100, 102, 104, 106, 108, 110, 112, 114, 115, 116, 117, 118, 119, 120, 126, 127, 128, 129, 130, 132, 134, 136, 138, 140, 143, 144, 148, 150, 152, 156, 158, 160, 161, 162, 164, 170, 172, 174, 177, 178, 180, 184, 185, 186, 190, 191, 194, 195, 199, 200, 201, 202, 204, 205, 208, 216, 217, 219, 220, 223, 224, 225, 231, 236, 240, 241, 244, 246, 250, 252, 264, 269, 270, 272, 273, 274, 276, 277, 278, 280, 281, 283, 284, 285, 316, 286, 288, 290, 292, 294, 296, 298, 300, 301, 302, 306, 308, 310, 312;

The Motoring Picture Library, National Motor Museum: 17, 21, 53, 57, 69, 73, 97, 101, 103, 133, 137, 141, 145, 165, 181, 198, 203, 214, 218, 230, 251, 255, 271, 291, 293, 317;

Pooks Motor Bookshop: 6, 7, 16, 82, 83, 84, 85, 92, 107, 154, 166, 168, 169, 182, 188, 189, 192, 193, 206, 207, 232, 234, 235, 238, 239, 248, 249, 256, 258, 259, 260, 261, 262, 266, 267, 282, 304, 305, 311, 314, 315;

Private collection: 146, 176, 210, 212, 222, 226, 242, 254.